The Space Shuttle

Other Books by Frank Ross, Jr.

ANTIQUE CAR MODELS:
Their Stories and How to Make Them

FLYING PAPER AIRPLANE MODELS

HISTORIC PLANE MODELS:
Their Stories and How to Make Them

HISTORIC RACING CAR MODELS:
Their Stories and How to Make Them

MODEL SATELLITES AND SPACECRAFT:
Their Stories and How to Make Them

CAR RACING AGAINST THE CLOCK:
The Story of the World Land Speed Record

JOBS IN MARINE SCIENCE

RACING CARS AND GREAT RACES

SPACE SCIENCE AND YOU

(COURTESY OF NASA)

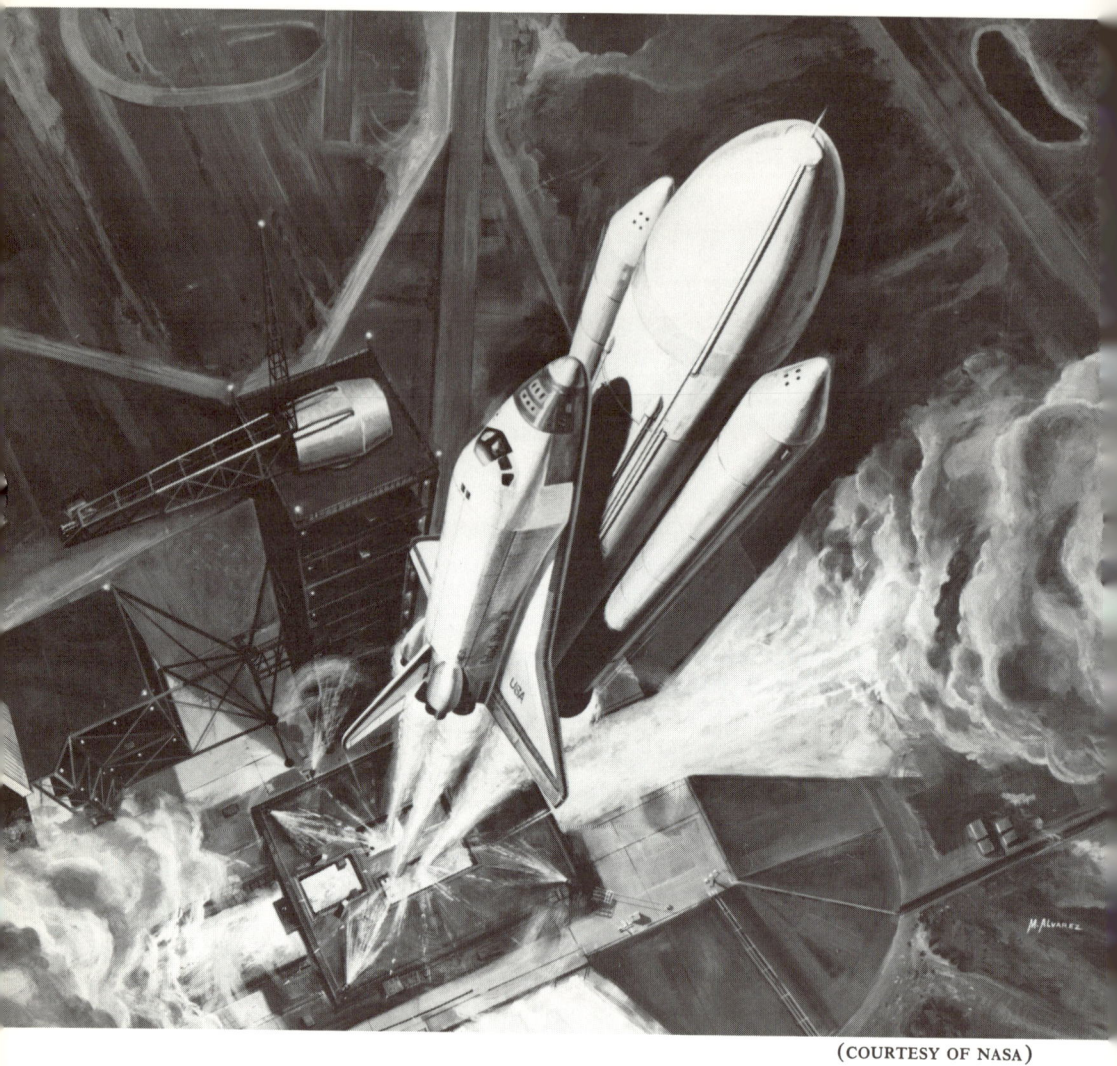

The Space Shuttle

Its Story and How to Make a Flying Paper Model

Frank Ross, Jr.
illustrated with photographs

Lothrop, Lee & Shepard Books
A Division of William Morrow & Company, Inc.
New York

Copyright © 1979 by Frank Ross, Jr.
All rights reserved. No part of this book may be reproduced or utilized in any form or by any means, electronic or mechanical, including photocopying, recording or by any information storage and retrieval system, without permission in writing from the Publisher. Inquiries should be addressed to Lothrop, Lee & Shepard Books, 105 Madison Ave., New York, N. Y. 10016.
Printed in the United States of America.
5 6 7 8 9 10

Library of Congress Cataloging in Publication Data

Ross, Frank Xavier, (date)
  The space shuttle, its story and how to make a flying paper model.

  Bibliography: p.
  Includes index.
  SUMMARY: An introduction to the space shuttle—its history, the construction of its major systems, a typical mission, and what it means in terms of future space travel. Includes instructions for making a simple flying paper model of the spacecraft.

  1. Reusable space vehicles—Juvenile literature. 2. Reusable space vehicles—Models—Juvenile literature. [1. Reusable space vehicles. 2. Reusable space vehicles—Models. 3. Models and modelmaking. 4. Space vehicles] I. Title.
TL795.5.R67   629.45′4   79-12155
ISBN 0-688-41882-1
ISBN 0-688-51882-6 lib. bdg.

FOR MASTER CARL NIGRO,
SUPER MODEL MAKER

CONTENTS

1.	History and Development	11
2.	The Shuttle Orbiter	27
3.	Rocket Engines and Orbiter Control Systems	37
4.	Power and Thermal Insulation Systems	46
5.	A Typical Space Shuttle Flight	53
6.	The Space Shuttle at Work	65
7.	Making and Flying the Orbiter Paper Model	77
	Space Shuttle Facts	83
	Milestones in Space Exploration	85
	Glossary	88
	Metric Conversion Table	91
	Further Reading List	92
	Index	93

(COURTESY OF ROCKWELL INTERNATIONAL'S SPACE DIVISION)

The Space Shuttle Enterprise is lifted on the back of its 747 carrier plane to a high altitude and then released.

1.
History and Development

The revolutionary space vehicle called the Space Shuttle is a combined spacecraft and airplane. It is launched like a rocket, riding a tower of fire straight up. Reaching a predetermined earth orbit, the Shuttle Orbiter maneuvers like a conventional spacecraft, using small rocket motors. At the end of its space mission, it reenters the earth's atmosphere to land at its ground base like an airplane. The Orbiter has no engine power at this point, so its landing is similar to that of a glider plane rather than that of a powered aircraft. After about two weeks of servicing, refueling, and loading a new cargo, the Space Shuttle will be ready to return to space on another mission.

Unlike the costly one-shot manned and unmanned missions of the last two decades, the Space Shuttle has been specifically designed to make repeated journeys into space and back to earth. With a working lifespan of about 100 missions, using almost all the same equipment, this reusable space vehicle promises to cut drastically the cost of future space programs.

During the pre-Shuttle years of space activities, sending a cargo, or *payload*, into *orbit* aboard an unmanned launch vehicle cost approximately $600 to $700 per pound. The planned fleet of five Shuttle Orbiters is expected to reduce this huge expense by as much as five times, to an estimated $100 per pound. This may fall even lower as Shuttle operations become more efficient. The Space Shuttle, scheduled for launch in late 1979, and its sister ships are the key units in a unique, pioneering Space Transportation System that marks the beginning of a new era of space activities.

The whole concept of the Space Shuttle and the revolutionary Space Transportation System it is establishing is not an overnight development. It represents more than three quarters of a century of scientific and technological achievements in heavier-than-air flight and, more recently, in spectacular journeys through space. Although the Space Shuttle is an extraordinarily modern vehicle in every respect, its heritage can nevertheless be traced back to the work of the Wright Brothers, who built and flew the first successful heavier-than-air flying machine in 1903. The crude controls they devised for their frail aircraft have been highly refined for the Orbiter's earth landing following its journey through space. The Orbiter incorporates many other airplane features, such as efficient wing design and landing gear, that slowly evolved over the years following the Wright Brothers' historic achievement.

The Space Shuttle owes some of its spacecraft features to even older technical developments. For example, the rocket power needed for launching the vehicle into orbit and for maneuvering it in space has a history of more than 2000 years. The Chinese are generally credited with inventing the rocket and using it in a simple form as a military weapon. Military rockets were continually used over the centuries, passing through various stages of advancement with increasing flying range and destructiveness. Written during the War of 1812, our national anthem makes mention of the "rocket's red glare."

Rockets for sending unmanned and manned spacecraft into orbit, however, are strictly twentieth-century technical developments. As with countless other scientific and technological accomplishments, contributions have been made by many experts from various countries. One of the greatest of the early rocket pioneers was Dr. Robert H. Goddard of the United States. He conceived and developed a number of basic features relating to rocket *propellants* and controls. He also proposed one of the first theoretical flights to the moon to be made by a rocket-propelled man-made vehicle.

Other great pioneers in the history of the modern rocket must include K. E. Tsiolkovsky of Russia, Dr. Eugen Sanger of Austria, and Dr. Hermann Oberth of Germany, all of whom developed theoretical designs for manned spacecraft journeys to the moon and elsewhere in space and, importantly, back to earth. Outstanding among the more recent rocket space experts was Dr. Wernher von Braun of Germany and later of the United States. It was von Braun and his scientific and engineering colleagues who, in Germany during World War II, designed and built the V-2, a fantastic rocket weapon of destruction. It traveled through the lower levels of space at a speed of more than 3,000 miles per hour, falling on its target without warning. There was no defense against it. The war ended in 1945 as von Braun and his engineering team were designing other, more advanced long-distance space rocket weapons.

Although primarily weapons of destruction, these rocket vehicles that von Braun and his associates created were in a real sense forerunners of the spectacular spacecraft launched by the United States and the Soviet Union in the 1950s and 1960s. At the end of the war, the mountain of technological information accumulated by the German experts was seized by both the United States and Russia. Later, a number of German rocket scientists and engineers, including von Braun, came to the United States to assist in America's fledgling rocket development program whose goal was to send a satellite into orbit around the earth.

In great secrecy, Russian rocket experts were working toward that same objective. They too were helped by captured German rocket data as well as by individual scientists and engineers. Another longer-range objective of the Russian rocket experts was to build a spacecraft for sending human beings into space. Their counterparts in the United States had similar programs in the study and development stages, but they were in no rush. American rocket experts felt that, before a human being was shot into space, they needed a great deal more knowledge about the technology of a manned spacecraft and about the effects of the space environment on the human body.

Some of the hundreds of engineering problems that needed solution before a manned spacecraft could be launched into orbit around the earth included the development of practical and enormously powerful rocket motors for accelerating unmanned or manned spacecraft fast enough to counter the force of gravity and to achieve earth orbit; the creation of metal alloys to withstand the tremendously high temperatures of rocket engines; the development of new heat-resistant materials for protecting astronauts and their spacecraft from burning up while returning through the atmosphere to earth. There were also highly complex problems concerning the control of spacecraft during launch, in orbit, and on the return to earth.

More important, however, were the as yet unanswered questions about the human factor. For example, could the human body stand the enormous stress of gravity and the tremendous speeds involved in space flight? Was man able to tolerate the airlessness and zero gravity of the space world, along with such hazards as ultraviolet radiation? The most efficient spacecraft designed and built for carrying human beings into space would be useless if space scientists discovered that man could not leave the earth-bound world.

Some solutions to both technological and human behavior problems were determined in the United States with the construction and flight of a large family of experimental supersonic rocket-propelled aircraft in the late 1940s and early 1950s. These included the X-1, X-2, D-558-2 Skyrocket, the X-15A, and the X-24B. It was in October, 1947 that the X-1, with United States Air Force Major Charles (Chuck) Yeager at the controls, answered the question of a human being's ability to survive supersonic speeds. In that historic flight, for the first time both man and machine flew faster than the speed of sound, or more than 700 miles per hour (1126 kilometers per hour). Thereafter, each succeeding and more advanced experimental plane was flown faster and higher. A spectacular altitude record was established in the early 1950s by Captain Milburn Apt of the United States Air Force when he piloted the X-2 to a height of 126,000 feet (38,404 meters), the lower reaches of space, while blazing along at a speed of 2148 miles per hour (3456 kilometers per hour).

The bits and pieces of engineering and human data produced by these and other research activities were eagerly studied by those involved with space projects. Enormously encouraged by the mass of new information, they became certain that space flight, unmanned or manned, was indeed possible. Then, suddenly, on October 4, 1957, the world was startled by the news flashed from Moscow that Russian space experts had successfully launched and orbited around the earth an artificial satellite, Sputnik I. A metal sphere 22.8 inches (58 centimeters) in diameter and weighing 184 pounds (83 kilograms), Sputnik I launched the space age.

The historic tiny Russian spacecraft initiated a long series of spectacular space events over the following two decades. Space scientists and engineers in the United States were greatly stimulated by the Russian achievement. Soon after Sputnik I, on January 31, 1958, America's first artificial satellite, Explorer I, was launched into orbit. Then on April 12, 1961, man successfully entered the space world for the first time when Russian cosmonaut Yuri Gagarin made a single orbit around the earth in Vostok I.

Literally hundreds of unmanned and manned spacecraft were subsequently rocketed into space from the United States and the Soviet Union. As the space vehicles became increasingly sophisticated, they performed a variety of jobs in space useful to people on earth. For example, satellites were designed and built for observing worldwide weather conditions; communication satellites were orbited for improving radio and television transmission between nations; and spacecraft were equipped with highly complex scientific equipment for collecting research data about the sun and such distant planets as Mars, Venus, and Jupiter.

The climax of all these early space activities began on July 16, 1969, when three American astronauts—Neil A. Armstrong, Edwin E. Aldrin, Jr., and Michael Collins—were launched aboard Apollo 11. After the spacecraft orbited the moon, Collins remained in the command vehicle that was to return the astronauts to earth while Armstrong and Aldrin climbed aboard a lunar module, separated from the command vehicle, and descended to the moon's surface. On July 20, 1969, the two astronauts left the lunar module and walked on the moon's terrain, accomplishing a dream of the ages. The American lunar landing was repeated six more times, the last occurring on December 11, 1972.

The mind-boggling space accomplishments of the pioneer years occupied a small army of scientists, engineers, and technicians who designed, built, and launched the varied fleet of unmanned and manned spacecraft. To carry out the many different kinds of space projects, huge sums of money were required. The Apollo 15 moon expedition, for example, cost approximately 500 million dollars. And much of the expensive space machinery was built for one-time use. One of the great weaknesses and costs of space activities at this time was the giant rocket launchers which, after placing a spacecraft in orbit, fell to destruction. The financial strain soon became overwhelming.

By the early 1970s, it became obvious that space activities could not continue on the same lavish scale. In the United States, a definite slowdown occurred. It was a time for evaluating what had been achieved in the past and for determining how the practical benefits of space could be realized in the future at a reasonable cost. If space flights were to be other than mere technical stunts, more practical and less expensive methods had to be found for operating them.

This rather critical state of affairs in space engineering initiated a number of studies that were aimed at finding more efficient space hardware, launchers, and spacecraft for the postpioneering era. Many scientists and engineers became deeply involved with this new space program. Their goal was to develop a safe, workable transportation system that would allow astronauts to journey into space and return to earth, using the same equipment over and over again.

In the United States, these space studies were under the direction of the National Aeronautics and Space Administration (NASA), which manages all of America's nonmilitary space projects. The ideas conceived by NASA's experts for a reusable spacecraft and even some of their design proposals were passed along to others in the aerospace industry, including Rockwell International's Space Division (builders of the Apollo spacecraft), Lockheed, and Boeing, who were expected to refine NASA's proposals and to present designs of their own. Two principal factors guided the development of these plans. One was that the future spacecraft had to be reusable; the second was that the cost of its development, construction, and operation had to be kept as low as possible.

Before the present design of the Space Shuttle was decided upon, a number of other, equally interesting proposals were suggested. Among these was a two-stage spacecraft consisting of a launch booster vehicle and an orbiter. Both units had wings jutting straight out from the body and both would be capable of returning to earth for reuse. Another similar concept called for the orbiter to be carried into space by a giant rocket booster vehicle; both orbiter and booster would have delta, or arrowhead-shaped, wings for flying back to earth. Still a third and extremely interesting design was a spacecraft called the Starclipper, the work of Lockheed aerospace engineers.

The Starclipper was to be a huge space vehicle (almost 200 feet or 61 meters long) in the shape of an arrowhead. Five powerful rocket engines in its tail would launch the spacecraft into orbit. Two giant tanks carrying the propellants for the rockets were to be attached along the edges of the wings and meet at a point at the front of the vehicle. Once the propellants were consumed, in a matter of minutes, the tanks were to be detached from the Starclipper, tumbling to destruction through the earth's atmosphere. After completing its space mission, the Starclipper would return to and land on earth, powered by a series of jet engines.

These early spacecraft designs were abandoned one by one due to such technical problems as structural materials required, availability of rocket engines, reusable capabilities, payload limitations, and the cost of construction and operation. Yet these design studies were not futile engineering exercises. A great deal was learned from them and many of their features were incorporated in the Space Shuttle and in the overall concept of the Space Transportation System. Sifting and analyzing all the various designs, NASA space experts decided in 1972 that the structural and operating features of the proposed Space Shuttle offered the best solution for future space travels.

Since many of the Space Shuttle's engineering ideas were the work of scientists and engineers of the Space Division of Rockwell International, that company was chosen by NASA in July, 1972, to go ahead with the final design and construction of the spacecraft. The main characteristics of the spaceplane had been firmly established. The reusable space vehicle, with a lifetime of at least ten years, would be capable of carrying a maximum cargo of 65,000 pounds (29,484 kilograms) into earth orbit. It would ride piggyback on a giant expendable propellant tank with its own three powerful rocket engines. In addition, it would have two auxiliary booster rocket engines that would be detached from the tank at a specific altitude and parachute back to earth. The boosters would be picked up by a recovery ship, then serviced, refueled, and used for another launch. The Orbiter itself would return and land on earth like an airplane. After going through a maintenance procedure, it too would be ready for another journey into space.

Although these design characteristics of the Space Shuttle and the entire concept of the Space Transportation System were decided upon, engineering refinements went on continuously. This was done not only to make the Space Shuttle a safer and better-performing spacecraft, but also to lower costs of construction and operation wherever possible. Even with economics in mind, the cost of developing the Space Transportation System is in excess of 6 billion dollars.

Rockwell International's Space Division was not alone in the extremely complicated task of building the Space Shuttle and the Orbiter's sister ships. Many companies throughout the United States, and some in Canada, were made partners in the huge engineering effort. They were assigned the construction of such components as the vertical tail fin, the landing gear, the reusable heat protective system for the Orbiter's entire outer body, the life support system within the crew-passenger compartment, and the Shuttle's communication equipment. As the farmed-out parts and systems were finished and tested, they were shipped to Rockwell International's plant at Palmdale, California, to be assembled.

Four years after Rockwell got the go-ahead to build the Space Shuttle, the first completed Orbiter model, OV-101 (Orbiter Vehicle), was rolled out of the factory on September 17, 1976. The first Orbiter was named the Enterprise by President Gerald Ford, after a spacecraft featured in a popular television series, "Star Trek."

Although the first of the fleet of Orbiters to be completed, the

(COURTESY OF NASA)

Orbiter Vehicle 101 (Enterprise) was rolled out of Rockwell International's Assembly facility, Palmdale, California, on September 17, 1976.

Enterprise was not scheduled to make the initial historic flight into space. The second Orbiter, OV-102, was to have that distinction. The Enterprise was strictly a test vehicle that was put through a long, exhaustive series of ground and flight tests to make certain the revolutionary spacecraft performed the way its engineers hoped it would.

The rigorous testing program to which the Enterprise was subjected was divided into four phases. The first involved a series of three unmanned ground taxi tests with the Enterprise attached to the back of a modified huge 747 commercial jet plane. The giant carrier aircraft had been altered and structurally strengthened to carry the spacecraft safely. The Orbiter was mounted piggyback on three pylons or struts that jutted up from the 747's body.

The ground runs began on February 15, 1977, and were mainly designed to find out how well the 747 could taxi with the heavy, awkward load on its back and if it would respond safely to steering and braking actions. These and other tests were performed at NASA's Dryden Flight Research Center at Edwards Air Force Base. This is a vast California desert complex of runways, hangars, and flight control and maintenance buildings, where new types of aircraft are put through strenuous flying tests.

These initial taxi runs proved successful, and the second phase of tests was begun on February 18, 1977. In this series, still without a crew aboard, the Orbiter was taken aloft on the back of the 747 carrier. These airborne trials had two main objectives. One was to provide engineers with information on how well the

The enormous size of the Orbiter Enterprise and its 747 carrier plane can be appreciated by comparing them to the human figures nearby.

(COURTESY OF ROCKWELL INTERNATIONAL'S SPACE DIVISION)

Enterprise's flight controls—*elevons,* rudder, and other devices (described in chapter three)—behaved as aerodynamic forces acted upon them. The second was to determine whether the 747 carrier could fly successfully with its massive load. It was important to know this, since the plan was eventually to use the big jet transport for carrying the Orbiter to various NASA-operated testing facilities in the United States and finally to Kennedy Space Center in Florida for launching.

The Enterprise came through the second test phase with flying colors. Preparations were soon under way for the third and more critical test phase—airborne flights. The Orbiter was taken aloft by the 747 carrier with a crew aboard who checked out a host of instruments. The first of these experimental air journeys took place on June 18, 1977, with two astronauts in the flight compartment, Commander Fred W. Haise, Jr. and Pilot C. Gordon Fullerton. The Enterprise rode piggyback on the 747 to an altitude of 14,970 feet (4562 meters) where the mated aircraft traced an oval pattern in the sky and then landed. Flight time was close to 56 minutes.

Two similar flights followed, the last on July 26, 1977. During these aerial trials, the crew members activated and checked the performance of a variety of the Orbiter's flight systems, including the electrical power units (fuel cells), auxiliary power devices, and hydraulic and coolant equipment.

After studying all the data which this short series of manned captive flights produced, engineers in charge of the test program declared the Enterprise ready for the next and even more critical tests—manned free flights. For these, the Orbiter was to be taken to a designated altitude by the 747 carrier and released. It would then glide back to earth, landing as it would after returning from a real journey into space. Commander Haise and Pilot Fullerton again were the crew for the first historic experimental flight on August 12, 1977.

Soaring slowly to a height of 24,100 feet (7345 meters) on

the back of the 747, the Enterprise was set free and began descending to the runway far below. Swooping earthward as though on an engineless glider, the astronauts lined their big spacecraft perfectly with the landing strip and touched down as they would in a regular airplane at a speed of 212 miles per hour (341 kilometers per hour). The flight earthward had taken 5 minutes 23 seconds. Stepping out of the space vehicle at the end of their nerve-tingling ride, Commander Haise described the Orbiter's flying ability as "superslick." Pilot Fullerton added his praise with the words, "a very stable airplane."

Four additional free flights were made, the last on October 26, 1977. Many factors involving the Orbiter's flying abilities had been proved out, but one of great importance showed that the giant spacecraft could stop rolling after touchdown within a three-mile (4.8-kilometer) runway. This is the length of the newly-built runways at Kennedy Space Center, Florida, and Vandenberg Air Force Base, California, where the Orbiters will land regularly after their trips into space.

The Orbiter Enterprise made a smooth landing on its second free flight test at Dryden Flight Research Center, September 13, 1977.

(COURTESY OF NASA)

This series of photos shows the Enterprise in various phases of a free flight test. In the photo on page 24, the Orbiter is carried aloft on the back of its 747 carrier plane. The photo on page 25 (top) shows the Orbiter and carrier at the moment of separation, while the photo on page 25 (bottom) shows the Enterprise heading for the runway at the far left.

(COURTESY OF NASA)

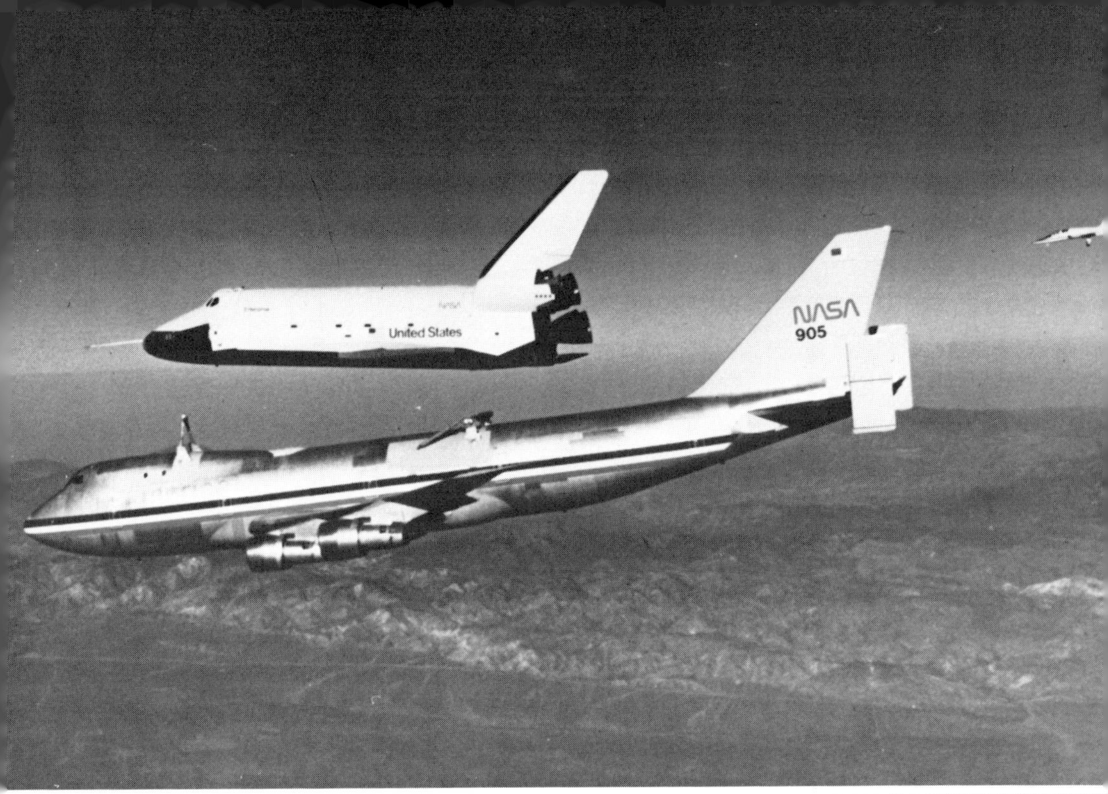
United States
NASA
905

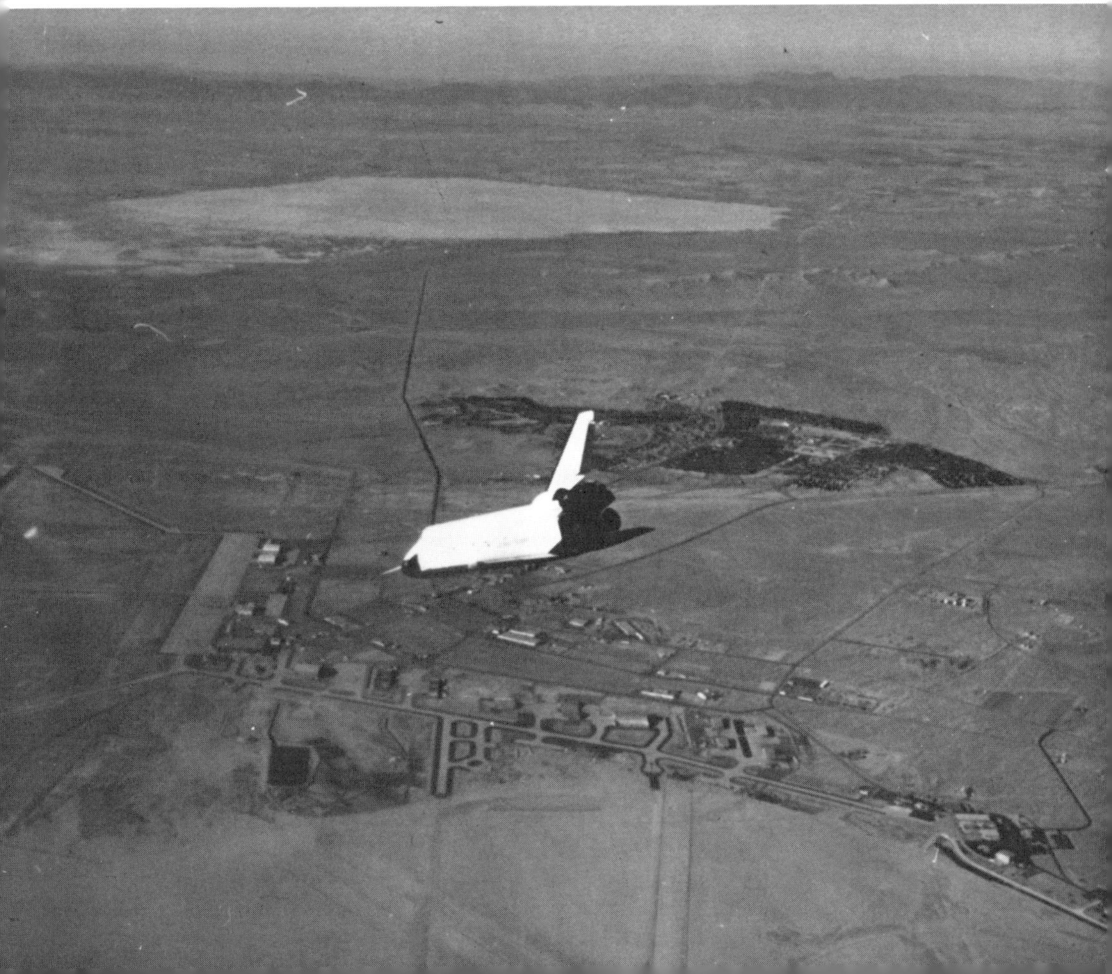

In March, 1978, the Enterprise was flown on the back of the 747 carrier cross-country to the Marshall Space Flight Center, Huntsville, Alabama, for its fourth and final series of tests. One of the remaining vital questions these tests answered was how well the structure of the Orbiter could withstand the enormous vibrations from its powerful rocket engines during launch. With the successful conclusion of the entire test program, the Orbiter was pronounced ready for the most crucial phase of its existence, flight into space.

Shuttle Orbiter OV-102, nicknamed the Columbia and the spacecraft designated for the historic launching, is scheduled to be completed and transported to Kennedy Space Center in the spring of 1979. A second crew of astronauts, Navy Captain John W. Young and Navy Commander Robert L. Crippen, was chosen to man the Orbiter on its pioneer voyage into space in late 1979. Those involved with the Orbiter's development are hopeful that after all the technical hurdles have been cleared, the revolutionary Space Transportation System will begin scheduled flights in 1980.

2.
The Shuttle Orbiter

The versatile Orbiter is huge, about as big as a DC-9 commercial jetliner which can carry more than 100 passengers. It has a 78-foot (24-meter) wingspan and a body length of 122 feet (37 meters). Aeronautical engineers call the Orbiter's wing design a double delta, after the fourth letter of the Greek alphabet. The wings extend from each side of the body at a sharp angle, then flare out at a broader angle to the wing tips, giving the entire Orbiter a giant arrowhead-like appearance. Without a cargo, the big spacecraft tips the scales at 150,000 pounds (68,040 kilograms), almost as much as fifteen mid-size automobiles.

The Orbiter's body is divided into three sections: a crew-passenger compartment at the forward end, a cargo bay, and a tail section housing the three main rocket engines and two orbital maneuvering rocket engines. A vertical tail fin and rudder tower rise above the tail end to a height of 27 feet (8.2 meters) from the top of the body.

The entire Shuttle Orbiter is made largely of aluminum alloy. Titanium, almost as light but stronger than aluminum, is used in a few places; in the rear section, for example, it helps to support the rocket engines. Finally, the outer surface of the spaceplane is covered with a variety of heat protective materials (described in chapter four) to prevent the Orbiter from burning up on its trip into space and the return journey to earth.

CREW-PASSENGER COMPARTMENT

The crew-passenger compartment and nerve center of the Orbiter is a three-level unit 2525 cubic feet (71.5 cubic meters) in size. The top level, or flight deck, is as big as a living room by comparison to the cramped quarters of the early manned spacecraft. It is airtight, as are the living quarters below, so the crew and passengers can work in an atmosphere closely resembling that on earth. Air in the cabin is a mixture of nitrogen and oxygen, kept to a sea-level pressure of 14.7 pounds to the square inch (1033.5 grams to the square centimeter.) This makes it possible for the crew and mission specialists to perform their duties without wearing bulky space suits with built-in life support systems. However, these space suits will be carried aboard the

This cutaway drawing of the right side of the Orbiter's crew-passenger compartment shows its three-tier layout.

(COURTESY OF NASA)

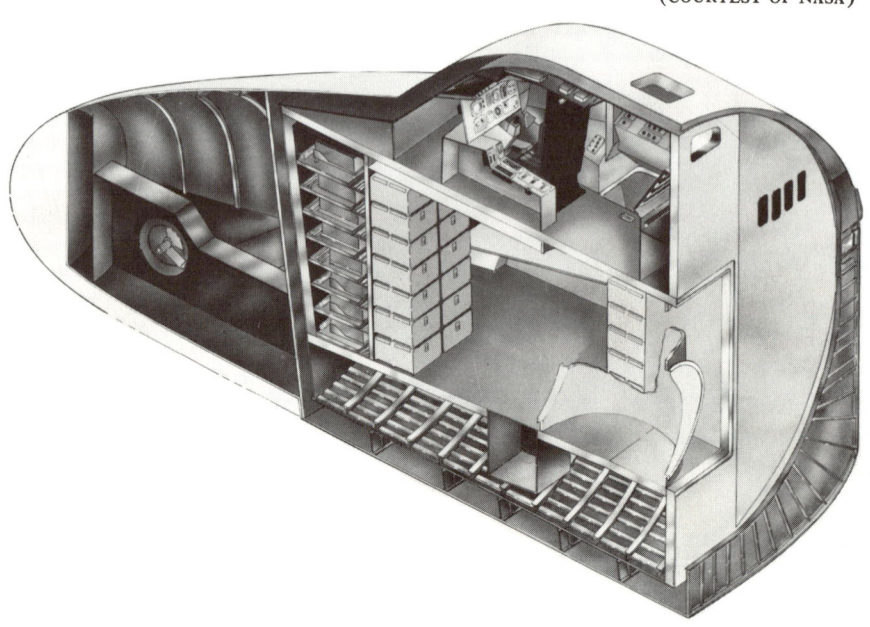

Orbiter and used by the astronauts when they must emerge from their orbiting spaceplane to check their vehicle or the cargo being *deployed* (released into space) or retrieved.

As on the flight deck of a commercial jetliner, two side-by-side, forward-facing seats are provided for the commander and pilot. On the forward wall of the flight deck and between the seats of the commander and pilot are dual instruments and controls for maneuvering the Orbiter during orbital flight and bringing it safely back to earth. Although operating the spacecraft is normally the task of the commander, this dual control design permits the vehicle to be operated from either seat and by only one person in case of an emergency. Above the instrument panels are a series of windows around the front of the

At the flight deck of the Enterprise with its dual instruments and controls are Fred W. Haise, Jr. (left) and C. Gordon Fullerton.

(COURTESY OF NASA)

flight deck which allow visual observation for the piloting crew.

Directly behind the commander is another seat that is occupied by a mission specialist. Depending upon the nature of the payload carried into space, the mission specialist may be a scientist or a highly-trained technician from NASA, a university, or private industry. Because of the near earthlike atmosphere in the cabin, he or she need not be a trained pilot astronaut—good physical health is the only requirement. Mission specialists are not involved with the operation of the Orbiter; instead, they are responsible for operating experiments and for deploying or retrieving the cargo. Two mission specialists can accompany the commander and pilot in the flight section of the Orbiter.

Once the Orbiter is whirling around the earth, the commander makes sure the spacecraft maintains its proper position while the pilot and mission specialist turn to the controls and instruments on the rear wall of the flight deck, called the payload

(COURTESY OF NASA)

The mission specialists will be seated behind the commander and pilot during launch.

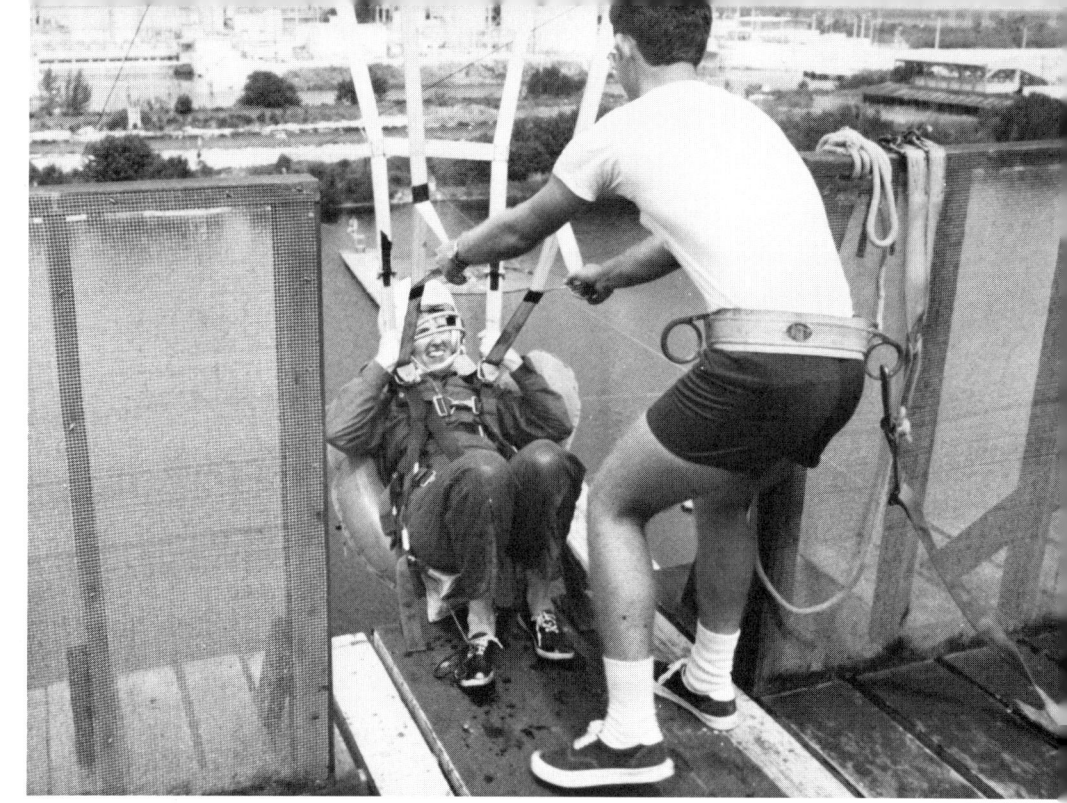

(COURTESY OF NASA)

Astronaut candidates undergo rigorous survival training. In this test, the trainee must be able to plunge into the water, release the parachute harness, swim under the floating parachute canopy, and climb into a life raft.

(COURTESY OF NASA)

A treadmill device tests the heart and lung condition of this astronaut applicant. Both astronauts and mission specialists must be in top physical condition.

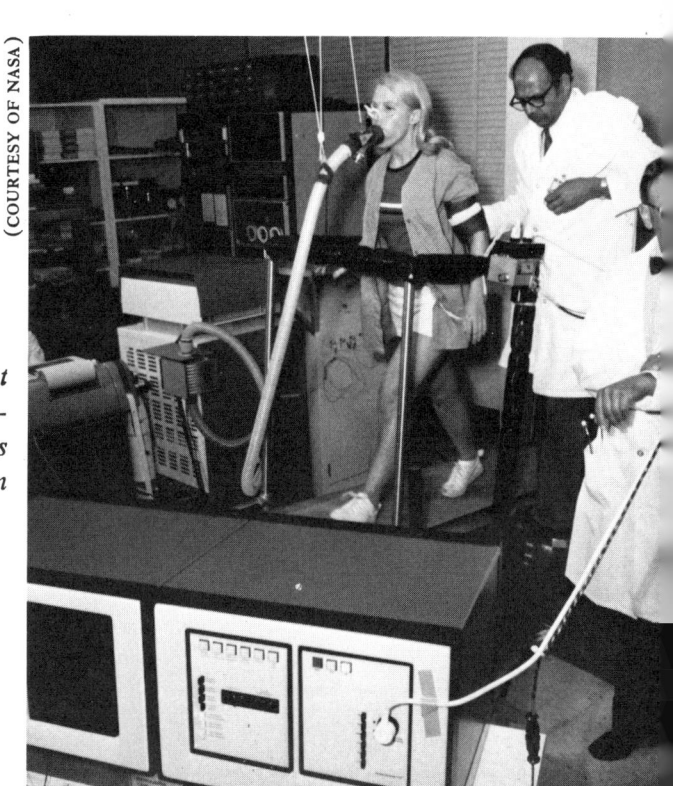

handling station. These instruments enable the mission specialist to monitor the cargo. If the cargo is a research satellite, for example, the mission specialist must make certain that it is properly serviced with such needs as electrical power and cooling fluids before it is deployed. Just prior to its removal from the Orbiter's cargo bay, the satellite's experiment-conducting instruments are activated from the payload station. Then, following deployment, the mission specialist keeps a constant check on the data being collected.

Other instruments and controls at the payload station permit the pilot to open the cargo bay covers and to operate the flexible mechanical arm in the bay used for removing the payload or retrieving it from space. To aid the pilot in cargo handling, small cameras in the cargo bay feed images to television display screens at the payload handling station. Two windows on the back wall of the flight deck also permit visual sighting of this operation.

Instrumentation and equipment for checking the Orbiter's subsystems, such as power supplies and communication apparatus, are located on the third and fourth sides of the flight deck. Some of this equipment is also used for servicing the payload.

The midsection, or middle deck, of the crew-passenger compartment is the living area. Here are bunks, a galley, oven, sanitation facilities, food and beverage storage compartments, water supply, and three hatches. A ceiling hatch provides access between the flight deck and living area; a side wall hatch permits an Orbiter crew member, wearing a life support space suit, to leave the spacecraft to check the outside of the vehicle in the event of an emergency. Also, if the Orbiter should dock to a space station or another spacecraft, the side hatch will allow access to it through an *air lock,* or connecting tunnel. A third hatch, located in the rear wall, makes it possible for an Orbiter crew member to crawl through a short tunnel into the cargo bay.

(COURTESY OF NASA)

The midsection of the crew-passenger compartment contains the Orbiter's kitchen facilities.

In the midsection too are three additional seats for extra mission specialists or, in case of emergency missions, astronauts rescued from a non-operating spacecraft. The three bunks in this section are simple flat pallets stacked one above the other against a side wall. They are easily removed and the space can be fitted with three more seats if necessary.

The galley is one of the more interesting units of the living area. Rising from floor to ceiling and about four feet wide, it is

33

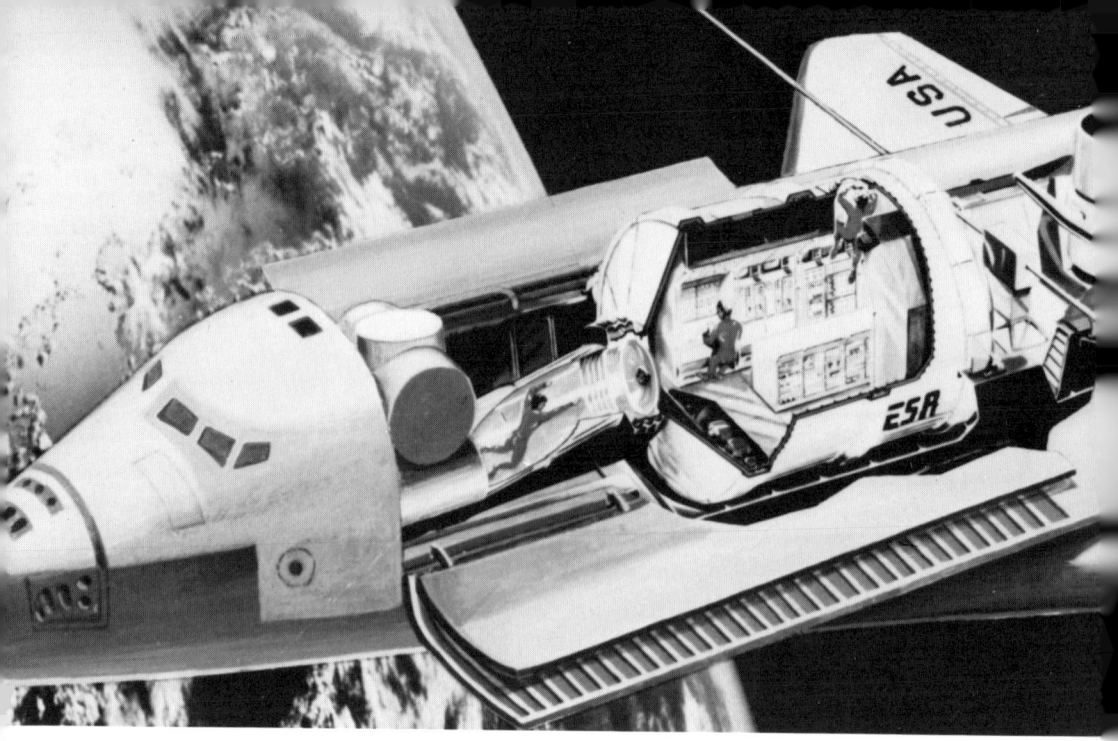

(COURTESY OF NASA)

This drawing shows an astronaut crawling through the air lock tunnel from the forward flight control section.

a model of storage efficiency. It contains bins for storing frozen and dried foods, ready-to-eat foods, emergency food supplies, drinking water, eating utensils, and trash as well as an oven for warming foods.

There are also storage compartments at the front end and in one corner of the midsection. These contain a variety of *avionics* equipment, largely electronic in nature, needed to operate the navigation, guidance, control, and communication instruments and devices on the flight deck.

The bottom or third section of the crew-passenger compartment is a cramped area occupied mostly by the Orbiter's environmental control equipment and storage lockers for the crew's equipment and clothing. The environmental control units are vital for maintaining the earthlike *pressurization,* comfortable temperature, and proper air mixture of nitrogen-oxygen and humidity of the crew-passenger compartment. Flight crew and ground service personnel can reach this area through covered openings in the floor of the midsection.

CARGO BAY

Directly in back of the crew-passenger compartment is the huge cargo bay. It is 60 feet (18.29 meters) long and 15 feet (4.57 meters) in diameter, the largest single section of the Orbiter. Designed to carry a payload weighing as much as 65,000 pounds (29,484 kilograms), it is roomy enough to accommodate a bus plus a mid-size automobile.

The extension or manipulator arm is the most interesting feature of the cargo bay. Attached to the left side of the bay looking forward and operated by remote control from the payload station, this flexible mechanical arm can extend as far as 50 feet (15 meters) to do its work. A television camera attached to the arm along with other cameras and lights strategically placed in the bay help the pilot via television screen to maneuver the arm.

Normally the cargo bay has only one manipulator arm. A second arm, however, can be added to the right side of the bay if required. This may be the case if the cargo is bulky or delicate and requires special care in handling. While the main job of the manipulator arm is to handle cargo, it may also be used by the flight crew to check the outer surface of the Orbiter. For example, if the crew should feel that some of the heat shield tiles (described in chapter four) are loose, the television camera on the manipulator arm can be used to examine them.

The top part of the cargo bay, also the upper surface of the Orbiter's body, is a set of movable doors. Operated from the payload station, these open in pairs to the right and left of the body after the Orbiter has reached its orbital destination so the cargo can be put aboard or deployed. The doors, of course, are closed during the periods of launch and landing.

TAIL SECTION

The tail section of the Orbiter is the business end of the spacecraft. The three powerful rocket engines that help boost the spacecraft away from earth are located here. This section of the Orbiter also houses two smaller rocket engines, one on each side of the tail, used for maneuvering into and out of orbit. There are, in addition, 44 miniature rockets on either side of the vertical tail fin; these, along with the others in the front end, allow the pilot to change the position, or attitude, of the Orbiter as it whirls around the earth. The Orbiter's various engines and other control systems are described in detail in the next chapter.

3.
Rocket Engines and Orbiter Control Systems

On launch, the Orbiter rides piggyback on a huge external propellant tank that houses the fuel for the three main booster engines in the Orbiter's tail section. The Orbiter receives its initial thrust from these main booster engines as well as from two solid rocket boosters attached to the sides of the external tank. Although the massive propellant tank is expendable, dropping away from the Shuttle as it nears the speed needed for orbit, the booster rocket engines, like the Orbiter itself, have been designed for reuse.

The Orbiter receives its final thrust into orbit from two orbital maneuvering engines. These engines along with 44 miniature rockets in the tail section are also used to adjust the spaceplane's position as it enters and during orbit and as it prepares for *reentry*. Following reentry, the Orbiter relies upon its airplane features in the wing and tail fin for maneuvering through the earth's atmosphere and for its glider-like landing.

Before describing each of these engines and control systems in detail, it is important to stress that none of the Orbiter's flight operations could be successfully performed without the help of a vast array of complex electronic instruments and computer-directed controls. While human hands will do much of the job of guiding the Shuttle into orbit and back to earth, some tasks will be done automatically through computers. However, if for any reason the automatic guidance system should fail, the commander or pilot can instantly take over the controls.

If a *rendezvous* with another spacecraft is needed, computer guidance will be used at the start of the maneuver. Actual docking to the space vehicle, however, will be accomplished manually by the pilot. Most of the computer guidance equipment is located in the nose end of the Orbiter and in the rear part of the flight deck.

MAIN BOOSTER ENGINES

The three main booster engines are positioned in a triangle pattern—one at the top and two side by side at the bottom—in the tail end of the Orbiter. Each of the rocket boosters is 14 feet (4.3 meters) long with a nozzle opening, the firing end, 8 feet (2.4 meters) in diameter. Roaring at maximum power

(COURTESY OF NASA)

Dummy engines are used to study the operation of the three main booster engines.

(COURTESY OF ROCKWELL INTERNATIONAL ROCKETDYNE DIVISION)

Technicians prepare one of the Orbiter's main booster engines for a test firing.

at lift-off, each rocket engine produces a *thrust* of 375,000 pounds (1,668,000 newtons). The rockets operate for about eight minutes from launch until the Shuttle almost enters its orbital path.

A unique feature of these three big booster engines is their ability to be moved over a limited range—up and down (pitch —10.5 degrees) and right and left (yaw—8.5 degrees)—as the Orbiter soars toward space. Called *gimbaling,* these engine movements permit the spaceplane's flight path to be altered if necessary.

(COURTESY OF NASA)

The external propellant tank is the largest component of the Space Shuttle.

EXTERNAL PROPELLANT TANK

The propellant tank is a giant cylindrical unit measuring 154.2 feet (47 meters) in length and 27.5 feet (8.38 meters) in diameter. Constructed of aluminum alloy, its main purpose is to carry the liquid oxygen (LO_2) and liquid hydrogen (LH_2) that propel the three main engines. The tank's interior is divided into two compartments to hold these chemicals separately. To protect this highly volatile unit from dangerous heat buildup during the first minutes of launch, the tank's critical surface points are coated with a silicone cork material and the remainder of its surface is sprayed with a coat of polyurethane foam insulation.

At lift-off, with both compartments filled, the propellants weigh 1,550,000 pounds (703,080 kilograms). The three flaming booster engines consume this great quantity in just about eight minutes and then shut down. At a predetermined altitude of about 69 nautical miles (122 kilometers), just before the Shuttle reaches the velocity needed for orbit, the empty propellant tank is blown free of the Orbiter by small explosive charges. This is accomplished remotely by the flight crew. Tumbling swiftly down through the earth's dense atmosphere, the huge, empty bulk glows fiery red, then breaks up into small pieces. When launched from Kennedy Space Center in Florida, these metal pieces will splash into a remote part of the Indian Ocean. On launches from Vandenberg Air Force Base in California, the hot shattered pieces of the tank will also fall into an empty region of the Indian Ocean.

SOLID-PROPELLANT BOOSTERS

In addition to the three main engines, two solid rocket boosters are used to give the Orbiter the speed needed to reach orbit. Attached to the sides of the huge propellant tank beneath the wings of the Orbiter (see Diagram 1), these rocket motors carry their propellants in solid form.

Long and slender, each rocket booster is 149.16 feet (45.46 meters) in length and 12.16 feet (3.70 meters) in diameter. They are fired at lift-off, along with the main rocket engines, and burn for about two minutes. The power each booster contributes at launch amounts to approximately 2,900,000 pounds of thrust (12,899,200 newtons).

About two minutes following lift-off and at an altitude of about 30 nautical miles (55.5 kilometers) and a speed of 3111 statute miles per hour (5005 kilometers per hour), the flight crew aboard the Shuttle ignites four tiny rocket motors at each end of the burned out booster motor, causing them to separate and fall away from the propellant tank. As the boosters fall swiftly back to earth, a small parachute (drogue chute) pops out of the nose end of each booster with the help of a timing

Diagram 1
Solid-Propellant Boosters

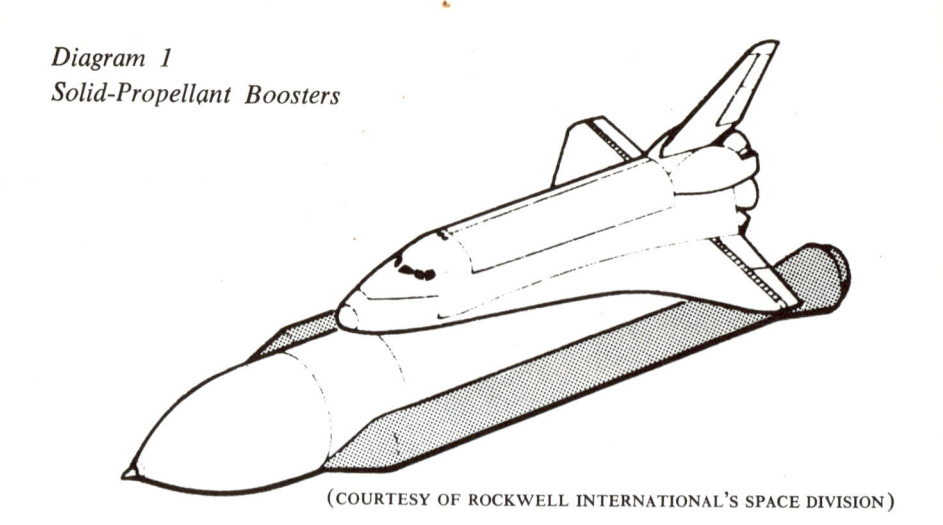

(COURTESY OF ROCKWELL INTERNATIONAL'S SPACE DIVISION)

device. This occurs at an approximate altitude of 19,000 feet (5,791 meters), and helps to stabilize and slow their descent. After dropping still further, to an altitude of 8,800 feet (2,682 meters), three larger parachutes break out of each booster and lower the spent rocket engines gently into the sea. The impact area is some 160 nautical miles (296 kilometers) downrange from the launch site. For Kennedy Space Center launches, the boosters will fall into the Atlantic Ocean; from Vandenberg Air Force Base, they will splash into the Pacific Ocean.

After they fall into the sea, a waiting recovery ship will hook a towline into each booster and tow it back to port. Returned to the Space Center or Air Force Base, the boosters will be reworked, recharged with fresh propellants, and made ready for another Shuttle launch.

ORBITAL MANEUVERING ENGINES

Soaring spaceward and seconds after dropping the giant propellant tank, the Orbiter's two orbital maneuvering engines are ignited. These are located topside on the tail end and on either side of the vertical tail fin. Each engine is covered with a streamlined shield or *pod*. The chief purpose of these smaller rocket engines is to give the space vehicle the final burst of speed needed to place the spacecraft into orbit, which may range from 100 to 600 nautical miles (185 to 1,111 kilometers) above the earth.

But the orbital maneuvering engines perform other tasks as well. They are fired when the piloting crew wishes to change the orbital path; to maneuver the Shuttle to a rendezvous with another spacecraft; and to carry out deorbiting maneuvers. In deorbiting, the engines slow the Orbiter's velocity, enabling it to glide down through the atmosphere to an eventual landing on earth.

Each of the orbital maneuvering engines develops a *vacuum thrust* of 6,000 pounds (26,688 newtons). The propellants used are monomethylhydrazine (MMH) as the fuel and nitrogen tetroxide (N_2O_4) as the *oxidizer*.

REACTION-CONTROL ENGINES

Since there are no air currents in space to create aerodynamic forces on the airplane features of the Orbiter—its wings, elevons, and rudder—to control its position in orbital flight, 44 miniature rocket engines are used. Strategically placed in the tail and nose sections of the Orbiter's body, they enable the flight crew to change the spacecraft's attitude in space, tilting it right or left and up or down, and to correct its attitude if necessary as it enters into orbit and during reentry.

These reaction-control engines consist of two types, primary and secondary, according to their size. There are 38 larger or primary engines and 6 smaller ones, all of which burn monomethylhydrazine (MMH) as the fuel and nitrogen tetroxide (N_2O_4) as the oxidizer. Fourteen of the primary control engines are located in the nose section of the spacecraft and 24 on the tail end, 12 on each side. The secondary control engines are known technically as *vernier* thrusters after their inventor, the sixteenth-century French mathematician Pierre Vernier. Of these six engines, two are positioned in the nose end and four are in the tail section, two on each side. The vernier thrusters produce only 25 pounds of vacuum thrust (111 newtons), and are used by the flight crew when small, precise attitude corrections are required. By contrast, the primary control engines develop 870 pounds (3,869 newtons) of thrust and are fired when more pronounced changes in the Orbiter's attitude are needed.

WING

In addition to its sophisticated rocket engines, the Shuttle Orbiter has two crucial airplane-like features, its double-delta wing and vertical tail fin. These are essential for navigating the spacecraft through the earth's atmosphere during its return to earth.

The wing's primary attitude control feature is movable panels called elevons, located along the rear edge of the wing. During a landing approach, the pilot can tilt these up or down in unison to correct the Orbiter's high or low glide position relative to the ground. The elevons may also be moved separately, one up and the other down, to help the rudder in the tail fin maintain the Orbiter's proper directional, or right and left, flight path as the vehicle drops closer to the runway.

VERTICAL TAIL FIN

The tall vertical tail fin rising from the topside of the tail section gives additional directional stability during the Shuttle's descent through the atmosphere and its landing approach. The tail fin's front edge angles sharply backward at 45 degrees, while its rear edge has a combined rudder/*speed brake* assembly. The rudder is divided into an upper and lower section and can be turned to the right or left, like the rudder on a boat. The pilot can control this movement as necessary for turning maneuvers. Each of the lower and upper rudder sections is split lengthwise in half. These are the speed brakes. The brake halves open in pairs like a book, and are operated by the pilot during the final landing phases. When the speed brakes are opened almost at right angles to the tail fin, they create a resistance, or drag, to the air currents flowing past the fin and help slow the spacecraft's speed.

Speed brakes are not a new device for slowing fast landing aircraft. They have been in use a number of years, particularly with the coming of jet planes and their extremely fast landing speeds. Speed brakes have proved most effective in safely controlling swift flying planes during landings and they are working in the same way for the Orbiter.

4.
Power and Thermal Insulation Systems

Without on-board power supply systems, the Orbiter would not be able to function as a spacecraft or to perform useful work while in orbit. And without adequate heat-resistant materials, the spacecraft and its crew would burn up during launch or reentry.

ELECTRICAL POWER

Electrical power on the Orbiter is obtained from fuel cells, electrical supply units about as big as storage batteries used in automobiles. On the Space Shuttle, a group of three cells is carried beneath the floor of the Orbiter's cargo bay toward its front end.

The first practical fuel cell for producing electricity was demonstrated by Sir William Grove in 1839. For more than 100 years, it remained a laboratory curiosity until the dawn of the space age in the late 1950s. Then engineers seeking a self-contained, lightweight power source for spacecraft rediscovered the fuel cell. Its efficiency was improved in order to produce greater power for a longer period of time. Fuel cells, along with solar cells, are now in common use as electrical power sources on manned and unmanned spacecraft.

Very simply, fuel cells produce electrical power directly from a chemical reaction. This results when a fuel, such as hydrogen, and an oxidant, oxygen, react in a chemical solution (an electrolyte) of potassium hydroxide. The fuel and oxidant are brought into contact with the potassium hydroxide through a porous anode tube (hydrogen) and a porous cathode tube (oxygen). The hydrogen and oxygen break down to release electrons, ions, and water. The electrons flow on through an electrical circuit to perform useful work. The ions travel from one electrode to another to complete an electrical circuit within the cell. The water is carried off as a waste product.

Enough chemicals are carried aboard the Space Shuttle to keep the fuel cells generating a total of 1,530 kilowatts for a normal seven-day mission. For thirty-day missions, more fuel is taken aboard. After returning to earth, the fuel cells are serviced and made ready for another mission. Each cell has a lifespan of 5,000 hours. Just a few of the many demands placed on the Orbiter's electrical supply system include operating the flight instruments, the instruments for monitoring the payload, the communications equipment, the television camera system, and the cooking oven in the living quarters.

HYDRAULIC POWER

The Shuttle's three separate hydraulic power systems are basically mechanical in nature: rods, levers, and other movable elements are set in motion by a special pressurized fluid. The fluid travels to the various movable elements through flexible tubing. On the Space Shuttle, this hydraulic fluid is pressurized by three pumps for the three individual hydraulic systems.

Hydraulic systems are used on the Space Shuttle where a large amount of power is needed for moving or controlling mechanical units. These units include the elevons, the rudder/speed brakes, landing gear, steering, and gimbal movements of the three main rocket engines.

THERMAL PROTECTION SYSTEM

The surfaces of the Orbiter experience tremendously high temperatures during launch and reentry through the earth's atmosphere. Heat buildup is most serious during reentry when the Orbiter, traveling at a speed of roughly 17,000 miles per hour (27,353 kilometers per hour), passes through increasingly dense layers of atmosphere, producing greater amounts of friction and heat. On launch, the problem is not nearly as serious, since the Orbiter flashes swiftly through ever thinner atmospheric layers, experiencing less friction and heat. In space, where there is no air as we have it on earth, there is little structural heat difficulty.

Analysis of wind tunnel tests on Shuttle models indicated three critical *thermal heat* areas of the Orbiter. According to these tests, the hottest temperatures would occur on the space vehicle's nose and leading edges of the wing, in excess of a scorching 2,300 degrees Fahrenheit (1,247° Celsius). During descent, these sections would approach white heat. The next and only slightly less serious heat problems would be on the underside of the wing and body and along the slanted front edge of the tail fin. Temperatures here would range between 1,200 and 2,300 degrees Fahrenheit (642° to 1,247° Celsius), giving these areas a cherry-red glow as the huge spacecraft descended swiftly through the atmosphere. Finally, it was determined that the temperature of the sides of the vehicle's body, sides of the tail fin, cargo bay covers, and upper wing surfaces would range between a mere 750 and 1,200 degrees Fahrenheit (395° and 642° Celsius).

Since the Orbiter is constructed largely of aluminum alloy, which begins to melt at a temperature of a little more than 1,200 degrees Fahrenheit (648° Celsius), engineers realized that the revolutionary spaceplane would have to be heavily protected with an effective thermal covering to prevent both the spacecraft and crew from becoming cinders during the return flight to earth.

Creating an efficient insulating material on such early manned spacecraft as the Mercury and Apollo had given engineers much knowledge about making special insulation for space vehicles. But these technical experts realized early in the design of the Orbiter that several unique features of the spaceplane meant that an entirely new, much more advanced thermal protection material had to be developed. Some of these features were the Orbiter's huge size, the extremely high temperatures to be encountered, and the fact that the insulating material had to be reusable, for 100 missions at least. On the pioneering manned spacecraft, protective insulation was good for only one flight, becoming scorched and blackened and losing its effectiveness.

In developing the Orbiter's insulating material, engineers turned to an abundant natural substance called silica, which is found in a variety of forms, such as quartz, in nearly all parts of the earth's surface. Ordinary sand is the best known form of silica, and glass is the most common of the many products in which it is used. One of its chief traits is high resistance to heat, making it an important ingredient in the manufacture of ceramic tiles used in industrial furnaces where super-high temperatures are encountered.

The thermal protective covering developed for the Orbiter consists of four main kinds. The first is a *reinforced carbon-carbon* material that protects the nose and leading edges of the wing where the highest temperatures occur. Those areas of the Orbiter with only slightly less critical temperature ranges—most of the under and upper wing surfaces, sides of the body, and sides of the vertical tail fin—are covered with high-temperature reusable insulation (HRSI) and low-temperature reusable insulation (LRSI). HRSI and LRSI are tiles made of silica quartz fibers compacted and fused into a solid form, then coated with a heat-proof ceramic layer. The white HRSI tiles are 6 inches (15.2 centimeters) square and the black LRSI tiles are 8 inches

(COURTESY OF VOUGHT CORPORATION)

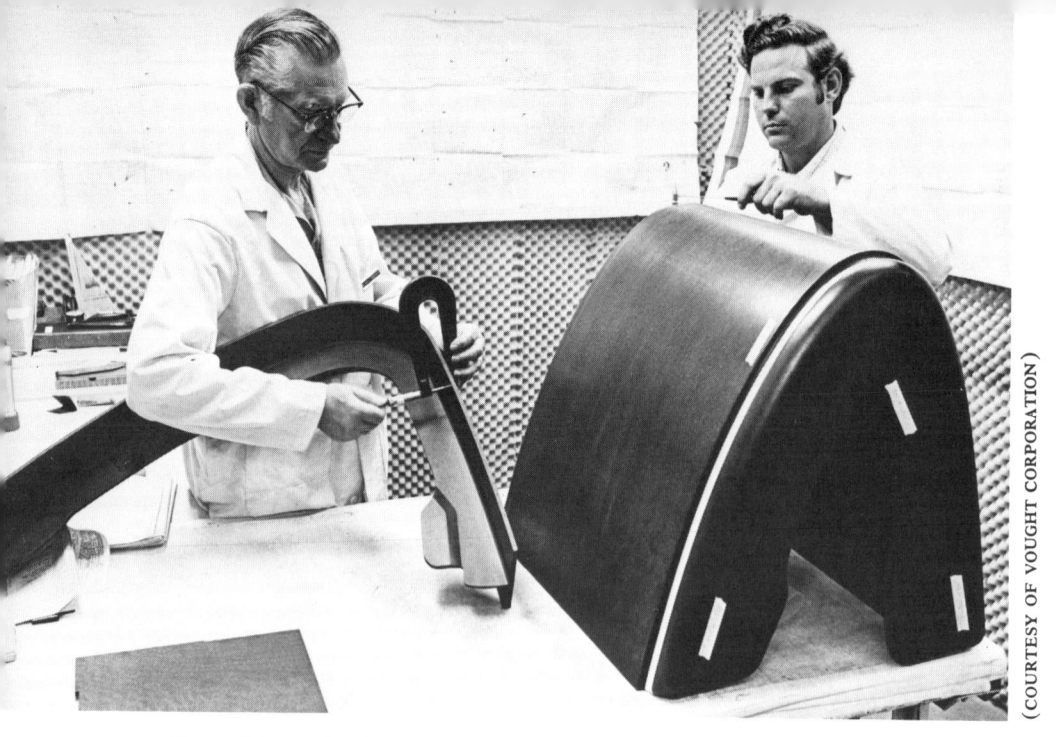

Technicians check one of the carbon-carbon panels to be bolted to the leading edge of the Orbiter's wing to protect it from intense heat during ascent and descent.

The insulated nose cap of the Orbiter is made ready for high temperature tests.

(COURTESY OF VOUGHT CORPORATION)

(20.3 centimeters) square. They differ only in the heat resistance of their ceramic surface. In tests, these newly-created tiles have shown a remarkable ability to protect against high temperatures. One side can be cool enough to touch with the bare hand while the other side is red hot. More than 34,000 tiles are precisely machined to fit the contours of the Orbiter, with no two alike. They are bonded to the Orbiter's outer surface with a special adhesive, giving it a somewhat checkered look. Finally, the least critical heat areas of the Orbiter are protected by a special insulating felt material with a silicone outer coating, known by its trade name of Nomex. This is used mostly on the cargo bay

The drawing shows the Orbiter with much of its surface white hot from heat buildup as it descends swiftly through the earth's atmosphere. Note the silica tile block insulation.

(COURTESY OF LOCKHEED MISSILE AND SPACE CO., INC.)

covers, a small area of the upper wing surface, and a small section of the sides of the body toward the rear.

The glass windows around the front of the flight deck are heat resistant because of the silica in their composition. They are able to withstand the scorching temperatures of the Shuttle's swift passage through the atmosphere during launch and reentry.

Laboratory tests have shown that with this cocoon of super heat resisting materials coating the Orbiter, the crew and their spacecraft will be protected from death and destruction during the fiery descent through the earth's atmosphere.

Diagram 2
Thermal Protection System

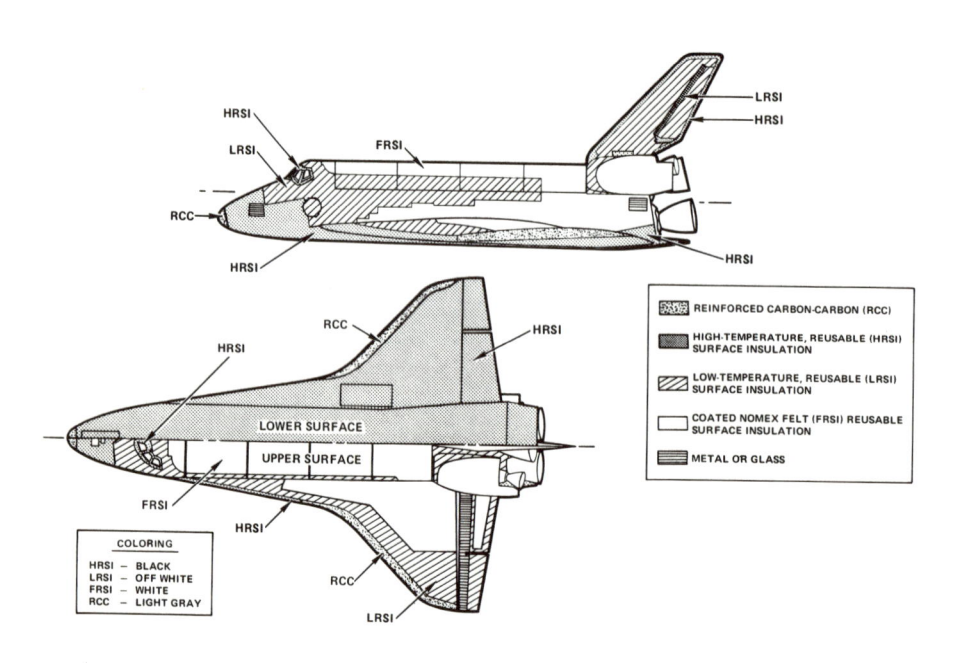

(COURTESY OF ROCKWELL INTERNATIONAL'S SPACE DIVISION)

5.
A Typical Space Shuttle Flight

Before launching, the Orbiter, external tank, and the two solid rocket boosters must be joined together as a single unit. The Orbiter is attached to the back of the giant propellant tank. The two booster rocket motors are fastened to the sides of the tank, beneath the wings of the spacecraft. During this assembling procedure all the units are raised to a vertical position. For the OV-102, all this will take place in the huge Vehicle Assembly Building at Kennedy Space Center.

A mammoth vehicle called the transporter then moves the assembled Space Shuttle to the *launch pad.* Here the payload is put aboard. The rudder and elevons are locked since they will not be needed until the Orbiter returns to earth. Then the flight crew and one or more mission specialists climb aboard the Orbiter.

At the command signal for lift-off, the solid rocket boosters and the Shuttle's three main rocket engines are ignited, rumbling and flaming furiously. Riding a spectacular column of fire and billowing smoke, the giant space vehicle rises slowly at first from the launch pad, then quickly picks up tremendous speed.

Approximately 30 nautical miles high (55.5 kilometers) and at a distance of 160 nautical miles (296 kilometers) downrange from the launch site, the two boosters, their work ended, separate from the Shuttle. Lowered gently into the Atlantic Ocean (for Kennedy Space Center launches) by parachutes, they are re-

This illustration shows the Orbiter being mated to the external propellant tank and solid rocket boosters inside the mammoth Vehicle Assembly Building at NASA's Kennedy Space Center.

This is an artist's rendering of the Space Shuttle on the launch pad as the moment for lift-off nears.

(COURTESY OF ROCKWELL INTERNATIONAL'S SPACE DIVISION)

(COURTESY OF ROCKWELL INTERNATIONAL'S SPACE DIVISION)

(COURTESY OF NASA)

The enormous Vehicle Assembly Building at Kennedy Space Center is the prime launch and recovery site for the Space Shuttle. The low building at the right is the Launch Control Center and the white strip in the upper left corner is the landing runway for the Orbiter on its return from space.

trieved by a waiting recovery ship, towed back to land, and serviced and recharged for another Shuttle launch.

Meanwhile, the Orbiter, still riding piggyback on the propellant tank, continues zooming at an ever faster speed toward space. Just before the spacecraft reaches the necessary *escape velocity* for orbit, approximately 17,500 miles per hour (28,158 kilometers per hour), the three main rocket engines stop firing and small explosive charges are set off, blowing the big tank free of the Orbiter. The huge empty unit falls back toward earth through the atmosphere, breaking up into glowing red-hot remains that splash in a remote part of the Indian Ocean. The Orbiter is now about 69 nautical miles (122 kilometers) above the earth.

The huge transporter, weighing almost six million pounds, is operated from two control cabins at opposite corners of the vehicle. It moves at a top speed of one mile an hour.

(COURTESY OF NASA)

(COURTESY OF ROCKWELL INTERNATIONAL'S SPACE DIVISION)

This is an artist's conception of the Space Shuttle, its main engines and solid rocket boosters roaring full blast, as it climbs through the atmosphere following launch.

(COURTESY OF ROCKWELL INTERNATIONAL'S SPACE DIVISION)

The huge external propellant tank is released shortly before the Space Shuttle reaches the escape velocity needed for orbit.

Soaring alone on the final lap of its journey to orbit, the Orbiter is given an extra burst of power by the orbital maneuvering engines to reach its goal. Depending on the type of mission the Shuttle is assigned to perform, its orbiting path may range anywhere from 100 to 600 nautical miles (182.5 to 1,095 kilometers) high and its stay in space from seven to thirty days.

Once in the desired orbit, preparations are begun to remove

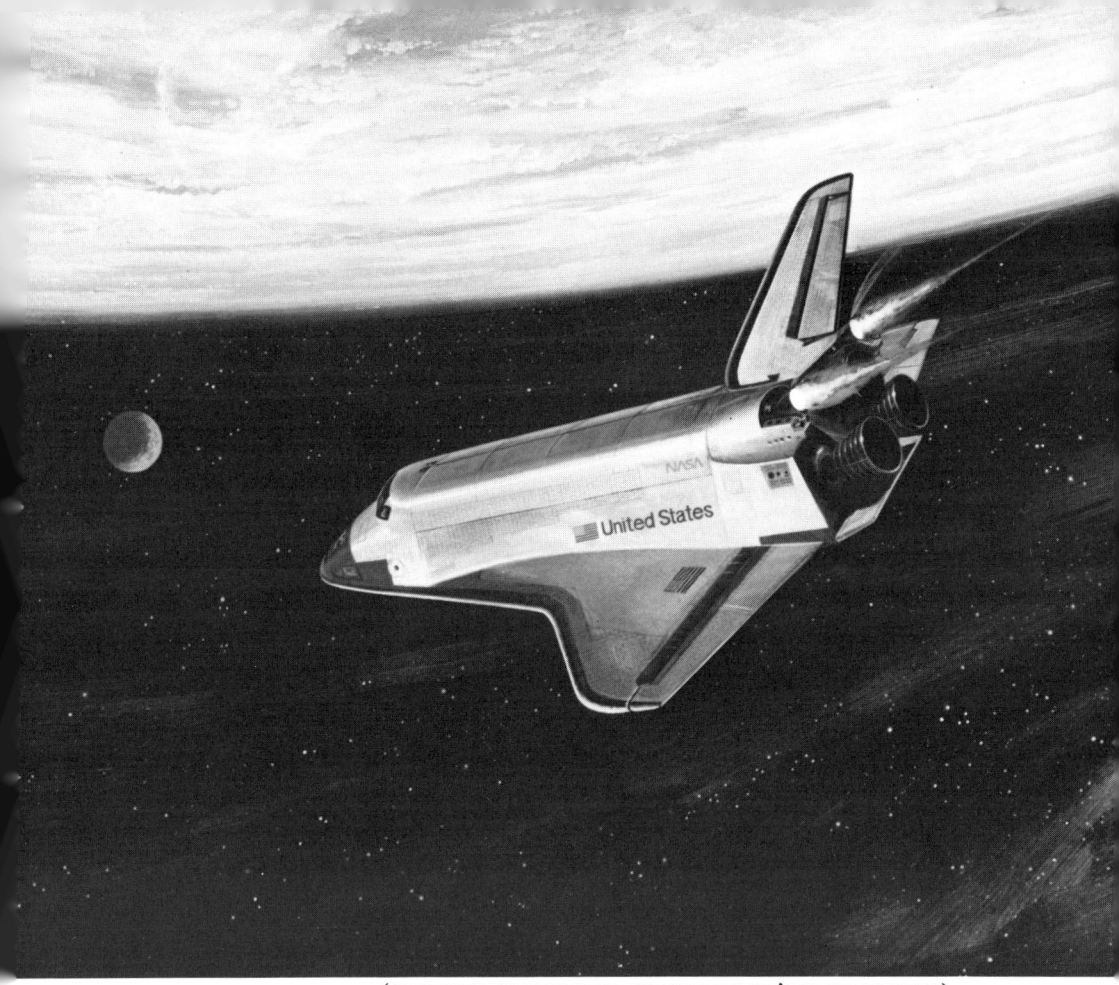

(COURTESY OF ROCKWELL INTERNATIONAL'S SPACE DIVISION)

Firing the orbital maneuvering rocket engines during deorbiting reduces the speed of the spacecraft.

the cargo. The pilot moves to the payload station on the flight deck to open the hatch covers on the cargo bay and to see that the manipulator arm is working properly for lifting the cargo out of the bay. The mission specialist checks the cargo to make sure it is receiving the services it needs for proper performance in space. Then he or she activates its instruments, and, following deployment, monitors their operation.

(COURTESY OF ROCKWELL INTERNATIONAL'S SPACE DIVISION)

This drawing shows the Orbiter about to reenter the earth's atmosphere.

Once the mission is completed, the commander and pilot check their instruments and controls in preparation for leaving orbit and returning to earth. The two orbital maneuvering engines are fired, initiating the reentry maneuver by reducing the spacecraft's speed. As the huge spaceplane descends into the upper fringe of the atmosphere, about 400,000 feet (121,920 meters), it is moving at a speed of more than 16,629 miles per hour (26,756 kilometers per hour) and is still on an earth-circling path. Sinking lower into the denser region of the atmosphere, the spacecraft's nose is angled upward and its underside begins to glow like a red-hot coal. At an altitude of about 80 nautical miles (148 kilometers) the thrusters and orbital rocket engines are shut off. Now that the Orbiter is in the atmosphere, air currents flow over the vehicle so that it is controlled by aerodynamic forces on the wing, elevons, tail fin, rudder, and speed brakes. The return landing maneuver is automatic with controls directed by computers. However, the automatic system can be instantly overridden by the pilot with manual controls.

Descending to a height of about 70,000 feet (21,336 meters) above the earth, its speed having slowly dropped to a little more than 1,000 miles per hour (1,609 kilometers per hour), the Orbiter begins the critical phase of its landing approach. The spacecraft has no power at this stage and is moving through the air like an engineless glider. Its forward momentum is sustained by the long, gliding descent. Without power, changes in the Orbiter's position in relation to the landing strip are limited to only a little more than 1,000 statute miles (1,609 kilometers) on either side of the runway.

The landing strip comes into view when the Orbiter is about 10,000 feet (3,048 meters) high and some five miles (8 kilometers) from the runway. By this time, the speed brakes on the vertical tail fin have been opened, greatly slowing the Orbiter's speed. A microwave radio beam flashed along a predetermined path picks up the spacecraft and aids in guiding it to the runway.

At a height of 2,000 feet (609 meters), the Orbiter is in a sharp nose-down angle. Now the flight crew lowers the landing gear as the Orbiter speeds home at 300 miles per hour (483 kilometers per hour). The speed brakes help reduce the speed to about 215 miles per hour (346 kilometers per hour) as the giant spacecraft finally touches down at the edge of the runway.

(COURTESY OF ROCKWELL INTERNATIONAL'S SPACE DIVISION)

This is an artist's conception of the Orbiter's landing approach at Kennedy Space Center. Vandenberg Air Force Base and Edwards Air Force Base in California provide two alternate touchdown facilities for the spacecraft.

At Kennedy Space Center and Vandenberg Air Force Base, the landing strip stretches for three miles but the Orbiter comes to a stop long before the runway's end as the flight crew applies the landing gear brakes. The landing speed of the Orbiter compares favorably with some of the jumbo commercial jet planes like the 747, which touches down at a rate slightly better than 160 miles per hour (257 kilometers per hour).

Immediately after the Orbiter lands and is towed to the servicing building, a highly trained ground crew takes charge of the spacecraft to prepare it for another mission. The normal servicing time expected is about two weeks. However, if there is an emergency need for the spacecraft to get back up into space—if, for example, astronauts orbiting in another Shuttle or Space Laboratory are in difficulty and need to be rescued—the landed Orbiter can be prepared for launch a good deal sooner.

The first servicing procedure after the Orbiter lands is called *safing*. This involves draining any remaining flammable or toxic fluids from fuel feed lines and removing unused explosive

This pilot's-eye view of the Space Shuttle's runway at Kennedy Space Center shows the landing lights being tested.

(COURTESY OF NASA)

charges. Then the entire aerospace vehicle is gone over with a fine-tooth comb, checking and servicing such things as the thermal protection system, the Orbiter rocket engines, the landing gear, the power systems, and flight instruments and controls.

After servicing and inspection, the Orbiter is towed to the Vehicle Assembly Building where it is mated to another external propellant tank. Two solid rocket boosters are also attached. Finally, riding on the transporter once again, the Shuttle is brought to the launch pad for another swift journey into space.

Most of the flights of the Space Shuttle will start from Kennedy Space Center in Florida. Launchings from this space port will send the Shuttle into east-west orbital paths. Other Space Shuttles will be launched into space from Vandenberg Air Force Base in California and will travel in north-south orbits.

All phases of a typical Space Shuttle mission, from launch to earth landing, are shown in this drawing.

(COURTESY OF NASA)

6.
The Space Shuttle at Work

The Space Shuttle has been designed and built to perform a variety of jobs in space. Its main function is to serve as a transport vehicle, carrying technically trained people and cargo into space and returning them to earth. As experience is gained in the operation of the Orbiter and as the practical uses of space increase, the functions of the Space Shuttle are also bound to expand.

Some of the jobs planned for the Shuttle in the immediate future involve transporting satellites into *low orbit,* 100 to 600 nautical miles (185 to 1111 kilometers) above the earth. The Shuttle is also capable of rendezvousing with a low-orbit satellite that is not working properly. If the trouble is minor, the satellite can be repaired in space by a technician wearing a space suit. If it is a "dead" satellite, the Orbiter crew can use the manipulator arm to retrieve it from space, taking it back to earth for repairs and servicing. This type of retrieval mission is expected to save millions of dollars in future satellite operations, since it will no longer be necessary to abandon non-working satellites, replacing them with new ones.

Another assignment planned for the Space Shuttle involves placing satellites in what is called geosynchronous orbit. From the earth, satellites in geosynchronous orbit, like those used for communications (television, telephone, and radio) and weather observation, appear to be stationary in space like a star. Actually, these satellites are spinning about 22,300 nautical miles (41,299

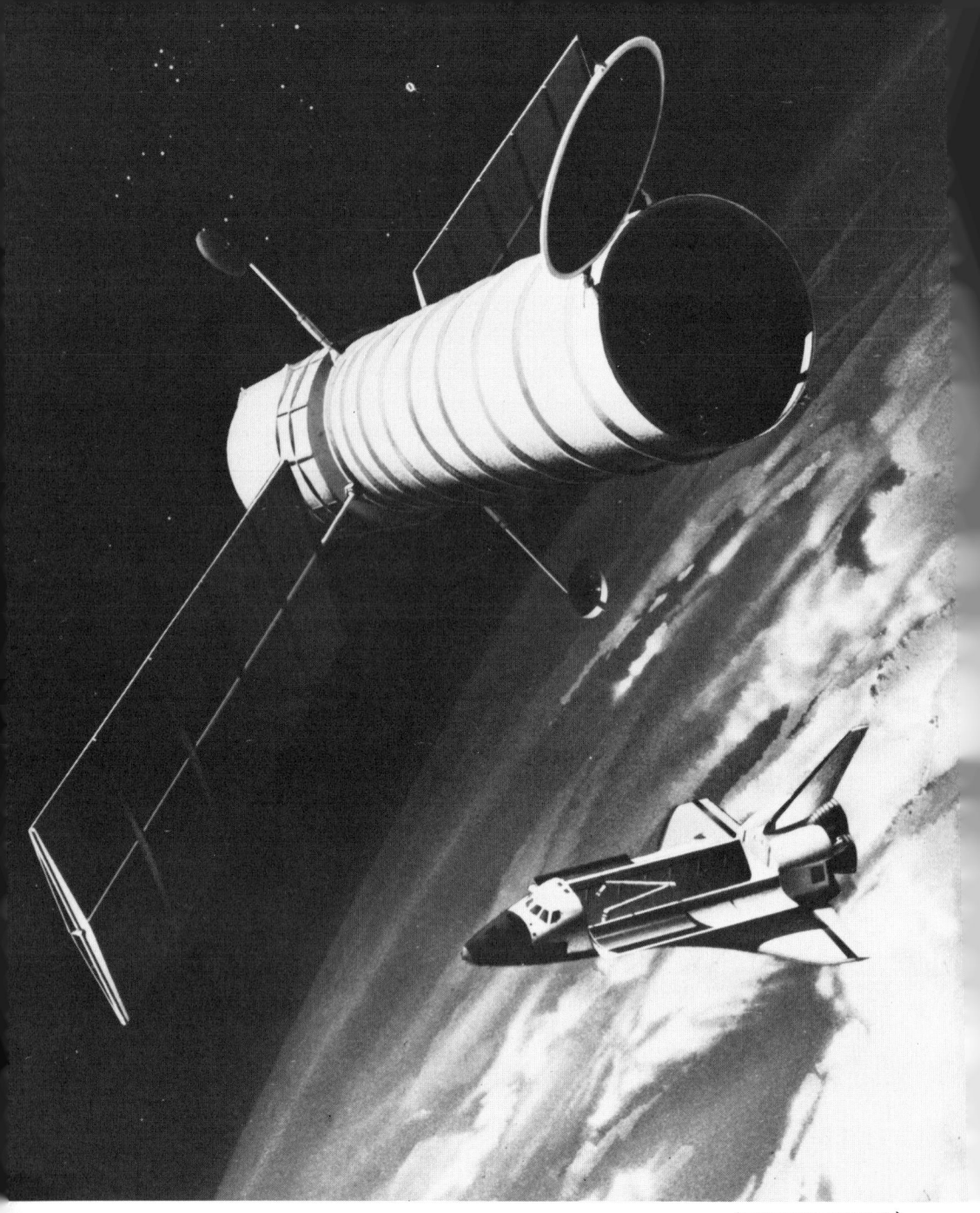

(COURTESY OF NASA)

One of the many jobs being planned for the Space Shuttle is carrying a large space telescope into orbit.

Technicians carried into space by the Shuttle Orbiter will be able to repair non-working satellites.

Payloads in orbit can be retrieved from space by the Space Shuttle's manipulator arm and brought back to earth for servicing and repair.

(COURTESY OF ROCKWELL INTERNATIONAL'S SPACE DIVISION)

Interplanetary-bound spacecraft will be carried to earth orbit by the Space Shuttle and then launched on their deep-space journey.

kilometers) from the earth at approximately the same speed as our planet.

From its low-orbiting path, the Orbiter will serve both as a launch base for these and other high-orbiting satellites and spacecraft. It will be assisted in this by a miniature reaction-powered vehicle known technically as a Teleoperator Retrieval System or, more commonly, "space tug." The small tug is equipped with two propulsion systems—one for long distance travel and the other for deorbiting and maneuvering purposes. Carried in the Orbiter's cargo bay, the tug will be placed in space by the manipulator arm and sent on its assigned task. This may be to boost a spacecraft into *high orbit* or to inspect, dock with, or retrieve a spacecraft that is in trouble.

The tug is remotely controlled by radio signals from the Orbiter. These are picked up by the tug's computer and translated into commands to the vehicle's various controls. When its

An artist's illustration shows the Teleoperator Retrieval system, or "space tug," on its way to help a spacecraft after deployment from the Space Shuttle.

(COURTESY OF MARTIN MARIETTA AEROSPACE)

job is finished, the tug can return to earth aboard the Orbiter or stay parked in orbit.

If the tug's job is to maneuver a satellite or other spacecraft into high orbit, both will be carried into space by the Orbiter. After deployment by the manipulator arm, the crew of the Orbiter will then move the spacecraft to a safe distance as the engine of the tug, started by a remote radio signal command, takes off for a predetermined long-distance destination. For example, if the satellite is for communications, it may be stationed over the Atlantic Ocean to link Europe and America or over the Pacific Ocean for communications between America and Asia.

Much the same technique will be used for retrieving a high-orbiting satellite (though not a geosynchronous satellite) that is no longer working. For this operation, the tug will be guided and controlled remotely by the crew aboard the Orbiter to bring the "dead" satellite to the spacecraft.

Using the Orbiter as a space base and transport vehicle for high-orbiting spacecraft is expected to save large sums of money. Launching high-orbiting satellites today costs as much as 40 million dollars; with the Space Shuttle the cost is expected to drop by more than half.

The procedure used for launching high-orbiting satellites is also being considered for sending unmanned spacecraft on interplanetary flights. One of the first missions of this type is planned for the spacecraft Galileo, scheduled for launching in 1982. The spacecraft's mission is to study the giant planet Jupiter and the space surrounding it.

Another future mission for the Space Shuttle involves carrying a space research unit called Spacelab into orbit. Spacelab is being designed and built under the direction of the European Space Agency, an organization formed by eleven western European nations. The lab is intended for scientific research. Half the unit will be equipped with a variety of instruments for

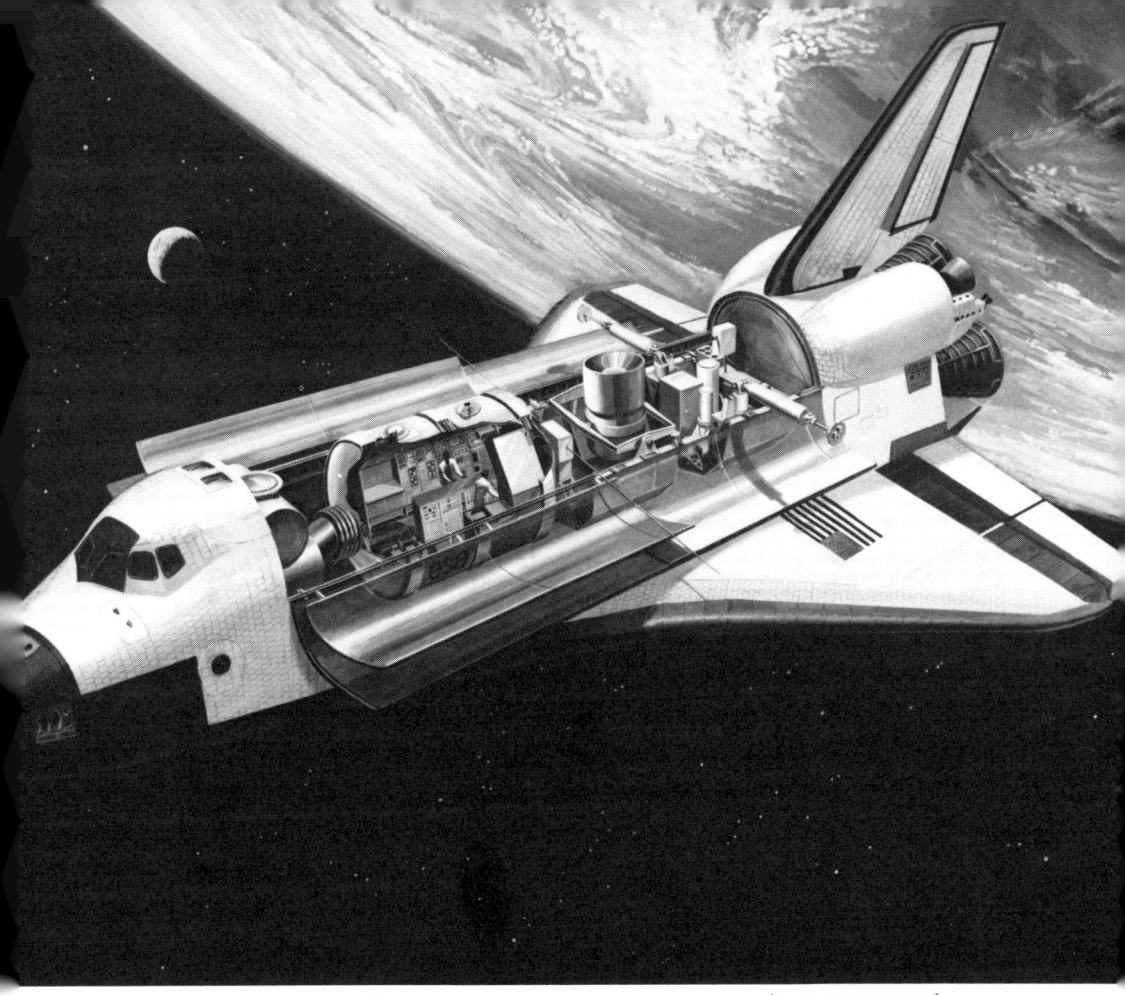

(COURTESY OF ROCKWELL INTERNATIONAL'S SPACE DIVISION)

This drawing shows the European Space Agency's Spacelab being carried into orbit by the Space Shuttle.

astronomical and geological studies and observation. The other half will consist of an enclosed, pressurized compartment where scientists and technicians, working in a shirt-sleeve environment, will monitor the data-gathering instruments and conduct experiments. A short tunnel connecting this section of Spacelab with the hatch in the rear wall of the Orbiter will allow the lab workers to move back and forth between the lab and the Orbiter's midsection.

Some of the experiments to be carried out on board the Spacelab will be in such areas as metallurgy (the creation of new alloys), medicine, and biology. A number of these were pioneered aboard America's first space station, Skylab, and are to be taken to more advanced stages by Spacelab workers. Skylab was launched into orbit on May 14, 1973, and was occupied by three different groups of astronauts for periods of 28, 59, and 84 days. The orbiting space station was last manned in 1974, after which time its systems were dormant and it began to slowly drop out of orbit. NASA had originally planned to rescue the space station by sending a manned Space Shuttle to attach a small rocket to it. Although Skylab's systems were revived and its position shifted, NASA abandoned plans to save the space station when it developed control problems and when delays in the rescue Shuttle voyage severely threatened the mission's success. The huge space station will now fall through the atmosphere to destruction sometime between mid-1979 and -1980.

Still other fascinating space tasks are being studied for the Space Shuttle. It may be used as a space truck and bus, bringing materials, supplies, and technicians into space for the construction and operation of solar power stations. This is not a dream. It is a practical, indeed, almost urgent goal made necessary by the energy shortage afflicting many areas on earth, especially the United States.

A number of concepts for space solar power stations are being studied. In general, these will be enormously large structures made of lightweight metals. Utilizing countless numbers of solar cells, these will absorb the sun's limitless energy and convert it into electric power. This in turn will be beamed toward earth by microwave transmission, picked up by earth receiving

(COURTESY OF ROCKWELL INTERNATIONAL'S SPACE DIVISION)

Carrying materials and technicians for the construction of giant solar power plants, as shown in this artist's drawing, is one of the tasks planned for future Space Shuttle missions.

stations, and then distributed through wires just like conventional electric power.

For the more distant years in the twenty-first century, the Space Shuttle concept may well serve as a transportation link between earth and space colonies. These man-made space

(COURTESY OF NASA)

This is an artist's conception of how a solar power satellite station, located some 36,000 miles above the earth, would beam its electrical energy to a land receiving facility for distribution to cities and towns.

islands would house thousands of inhabitants, living and working in an artificial environment.

Much of the technology already exists for the construction of these visionary space worlds. A leading proponent of this exciting space concept is Dr. G. K. O'Neill of Princeton University. He has suggested several possible designs for a space settlement. One would consist of two cylinders 20 miles (32 kilometers) long and over 21,000 feet (3378 kilometers) in diameter. Facilities for living and working in the space colony would be available for 200,000 to several million inhabitants. As seen by Dr. O'Neill, the space colony would be constructed of material obtained from the moon or from clusters of asteroids. These would be hauled to the construction site, possibly midway between the earth and the moon, by Space Shuttles that would be capable of soaring for greater distances from the earth than those now going into operation.

Life in this man-made space settlement would not be much different from that on earth. There would be comfortable housing, factories for work, and play facilities. Trees and grass would relieve the starkness of the surrounding space world, along with walks and bicycle paths winding their way through the greenery. There would be solar power stations to supply all the space colony's energy needs, and to create the accustomed sense of gravitational pull, each of the colony's huge cylinders would make a complete revolution every 114 seconds. Jobs at the factories would include manufacturing such products as exotic metals, ultrapure drugs and medicines, and electronic parts. These would be exported to earth markets. Food supplies would be obtained from the space colony's own farms, located in small, cup-shaped structures attached to the revolving colony. To visit relatives and friends or to take vacations on earth, the space colonists would ride regularly scheduled Shuttle flights. They would board a Shuttle, or disembark, at special airlock terminals built into the space settlement.

All this may sound like space fiction. But not too many years ago, when a similar fantastic proposal was made about the possibility of man traveling to and walking on the moon, it was received with the same skepticism. Now that event has actually occurred, not once but several times. So with the futuristic space colonies. They will come to pass as humankind feels the need to move on from a crowded, cluttered planet—perhaps not in the immediate future but surely in the lifetime of today's youth.

7.
Making and Flying the Orbiter Paper Model

MATERIALS AND TOOLS

White Paper—Weight is important here. The paper should not be too heavy or the model will not fly; if it is too light, the wings and body will not hold their shape. Two-ply drawing paper is excellent. Two sheets are all you'll need.

Pencil—for transferring the patterns to the white paper. A medium soft (#2½) is fine.

Ruler—a 12-inch or its metric equivalent.

Scissors—a strong, sharp pair for cutting out the patterns.

Knife—A sharp-pointed knife, such as an X-acto knife, is handy for scoring and for cutting out small corners of the patterns where the scissors will not do as well.

Draftsman's curve—for accurately drawing the curves. This is a plastic tool whose outer edges are shaped in a variety of curves. It is generally inexpensive and can be bought at an art supply store or a well-stocked stationery store.

Drawing pen—A pen using black ink is necessary for drawing the designs on the wing and body patterns. Many new types of pens are available with an automatic supply of ink and are perfect for this purpose. Most of them are inexpensive. Make sure when purchasing yours that the ink it uses is permanent. Niji Stylist is a popular brand and good for making these designs.

Glue—Both Elmer's Glue-All and Sobo glue are quick-drying and make a tight bond. A 4-ounce bottle of either brand is more than enough to complete your model.

GENERAL DIRECTIONS

1. Patterns for the wing and body of the paper model Orbiter are included in this book. Use either the United States measurements or the metric measurements, depending on which kind of ruler you are using. With your pencil, draw the outlines of the wing and body pieces on the white paper as accurately as you can. Use the draftsman's curve for the more difficult curves. The accuracy with which you do this will not only make your finished model look good, it will also help its flying capability.
2. Before attaching the body to the wing, draw all the pattern designs in pencil, then in black ink. Draw the designs as carefully as possible since these will also add to the attractiveness of the finished model.
3. When attaching the body to the wing, be sure that the two parts are at exact right angles to one another; otherwise the model will not fly properly. Also, if the parts are crooked, this will detract from the finished model's appearance.

MAKING THE PAPER MODEL

Wing

Draw and cut out the wing as shown in the pattern. Begin drawing the pattern with a center line. This is helpful not only for making one half of the wing exactly like the other, but it will also serve as a guide for gluing the body to the wing. Draw the designs on only one side of the wing, first with pencil, then with black ink.

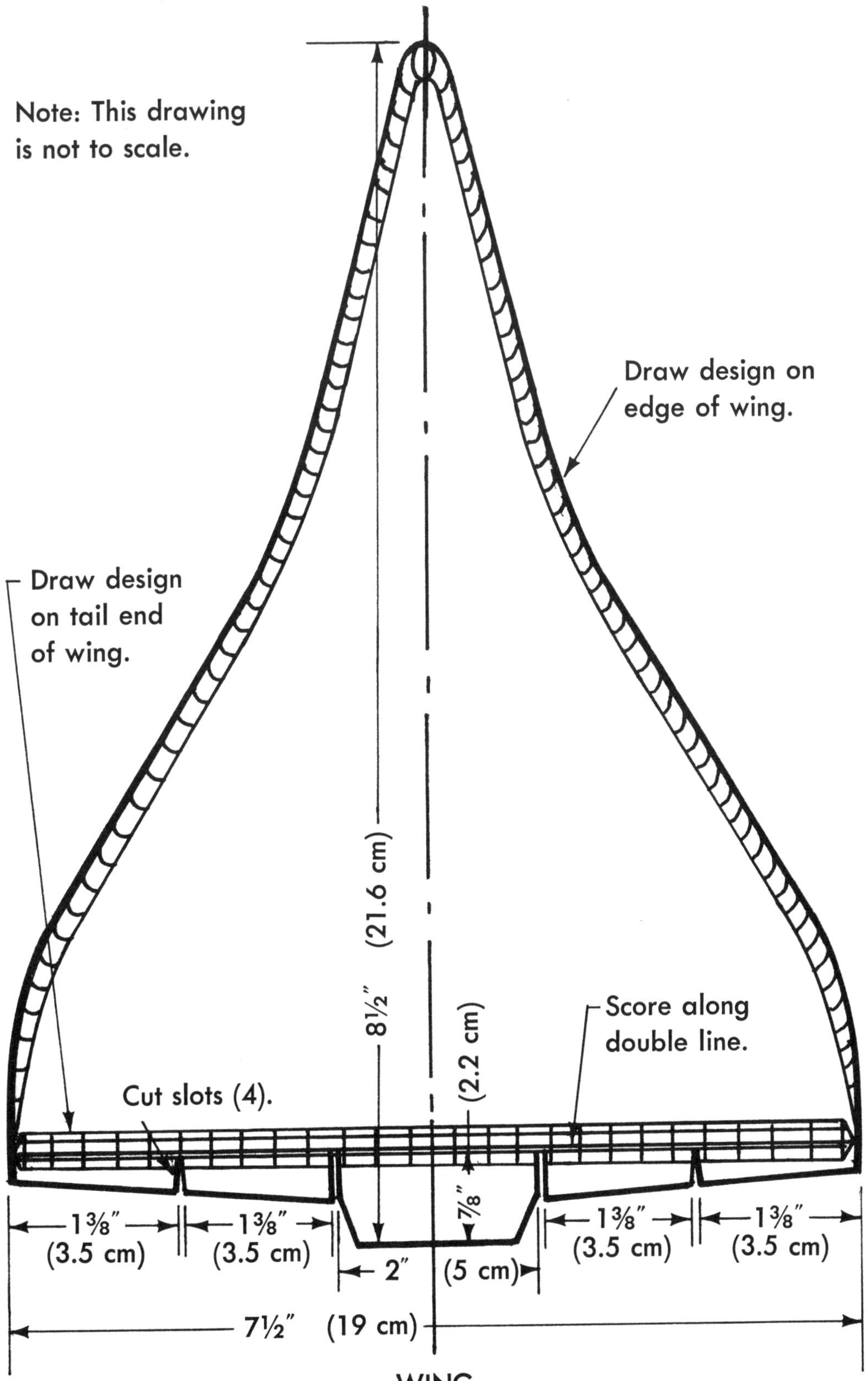
Note: This drawing is not to scale.
Draw design on edge of wing.
Draw design on tail end of wing.
8½″ (21.6 cm)
Score along double line.
(2.2 cm)
Cut slots (4).
7/8″
1⅜″ (3.5 cm)
1⅜″ (3.5 cm)
2″ (5 cm)
1⅜″ (3.5 cm)
1⅜″ (3.5 cm)
7½″ (19 cm)

WING

Using the scissors, cut the four slots on the rear edge of the wing as indicated on the pattern. With the wing lying perfectly flat on your work surface, use the X-acto knife to score lightly along the double line near the rear edge of the wing as shown on the pattern. The scored line together with the slots will allow the elevons to be bent upward and downward.

Body

Draw and cut out the body as shown in the pattern. Its outline looks more complicated than it really is. As with the wing, you will find the draftsman's curve a big help. Cut the body from the white paper, using the X-acto knife for getting into the small, tight corners.

Draw the body designs on both sides of the pattern. Do this first with pencil, so you can correct any errors, then with the black ink drawing pen.

Use the scissors or the X-acto knife to cut the slots on the vertical tail fin. Then, with the X-acto knife, score lightly along the heavy line of the tail fin as shown in the pattern. The knife is sharp and can easily cut through the paper, so be very careful applying pressure when you score. Also, make certain that a hard surface, a piece of wood or other material, is under the body as you score.

Now you are ready to attach the two pieces. Place the wing perfectly flat on your work surface. Position the body piece exactly along the center line drawn on the topside of the wing. Put a generous amount of glue along the center line. Wait a few minutes until the glue becomes tacky, then press the bottom edge of the body onto the center line. To avoid holding the body with your hands until the glue dries, put small objects— such as an ink bottle, paint jar, or paperclip box—against it on both sides. Check once more to see that the body is in a perfectly

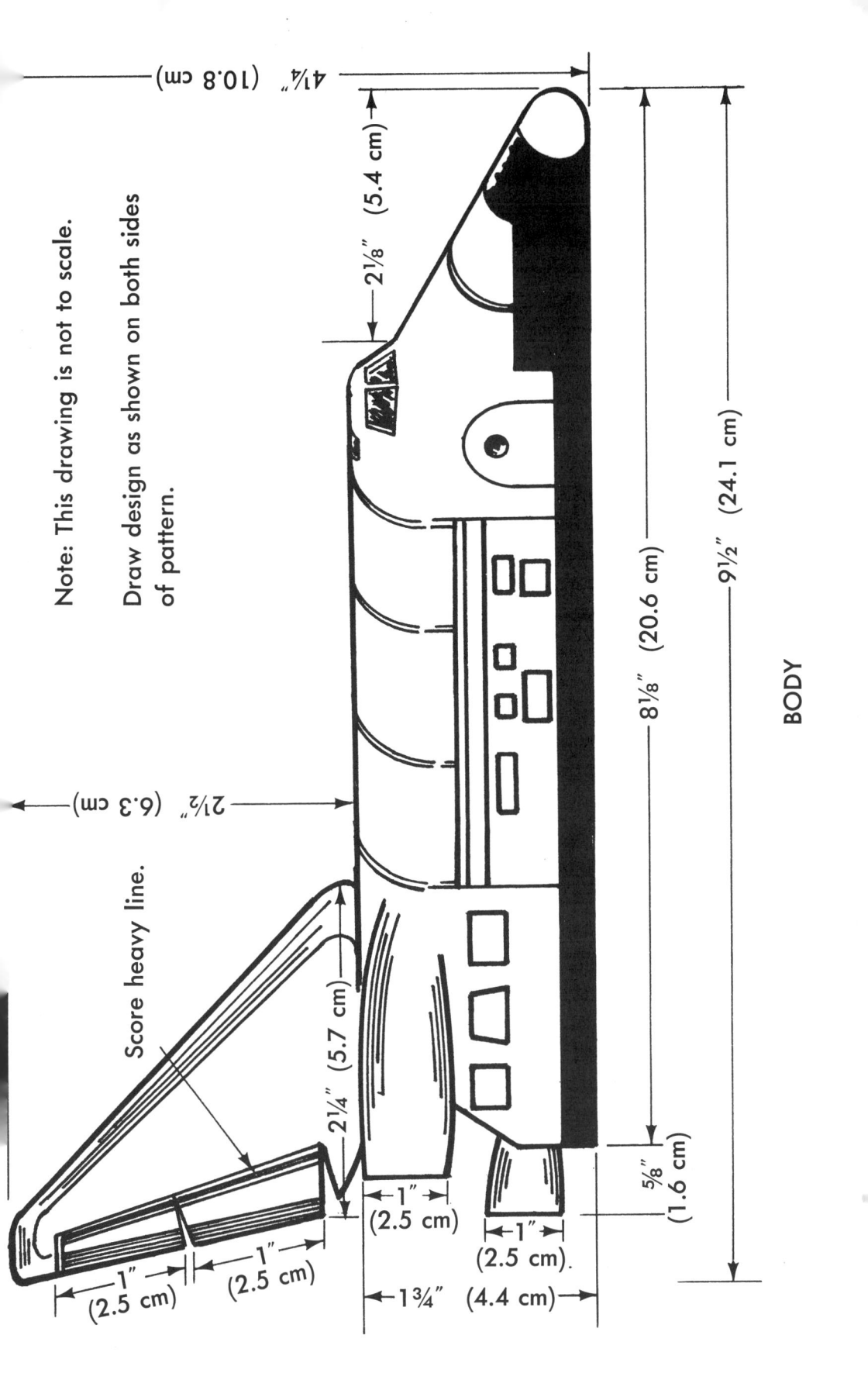

BODY

Note: This drawing is not to scale.

Draw design as shown on both sides of pattern.

Score heavy line.

4¼" (10.8 cm)
2⅛" (5.4 cm)
2½" (6.3 cm)
2¼" (5.7 cm)
8⅛" (20.6 cm)
9½" (24.1 cm)
5⁄8" (1.6 cm)
1" (2.5 cm)
1" (2.5 cm)
1" (2.5 cm)
1" (2.5 cm)
1¾" (4.4 cm)

upright position and exactly along the center line. Set the model aside and allow the glue to dry for at least 20 minutes.

FLYING THE PAPER MODEL

Of course, the paper model Orbiter cannot be launched as the real spacecraft will be—straight up like a rocket. But it can be made to land in a similar swooping glide and this can be fun.

To send the paper model Orbiter into the air, do not throw it as you would a stone. Hold the model near the front end or wherever you find it most comfortable and with a gentle overhand swing of your arm, let it go. Aim it upward rather than level since this will make it fly farther.

Once you perfect the launching, you can experiment with a variety of landings. You can do this by bending the elevons up or down, and by bending the rudder in the tail to the right or left. Bending the elevons up will cause the model to follow a rising path in the air after launching. Twisting them down will make the Orbiter model glide downward faster than during a normal launch. Also, by bending one set of elevons upward and the other downward, you can make the model tilt in flight. The tilt will be to the right or left depending on the alternate up or down position of the elevons. Bending the rudder to the right or left will make the model glide to the right or left.

You may have a problem with balance. After launching, the paper model may pitch sharply downward at the nose or drop sharply at the tail. This can be corrected by attaching a paper clip either on the nose section or the tail end. Trial and error will tell you exactly where the clip is to be attached.

Practice flying will soon make you a paper model Space Shuttle astronaut!

Space Shuttle Facts

(Values are approximate)

LENGTH
    Orbiter: 122 feet (37 meters)
    Space Shuttle: 184 feet (56 meters)
    Cargo bay: 60 feet (18 meters)

HEIGHT
    Orbiter: 57 feet (17 meters)
    Space Shuttle: 76 feet (23 meters)
    Cargo bay: 15 feet in diameter (5 meters)

WINGSPAN
    Orbiter: 78 feet (24 meters)

WEIGHT
    Gross lift-off: 4,500,000 lbs. (2,000,000 kilograms)
    Orbiter landing: 187,000 lbs. (85,000 kilograms)

THRUST
    Solid rocket boosters (2): 2.9 million lbs. of thrust each at sea level (12,899,200 newtons)
    Orbiter main engines (3): 375,000 lbs. of thrust each at sea level (1,668,000 newtons)

SEPARATION OF SOLID PROPELLANT BOOSTERS
    Height: 30 nautical miles (55.5 kilometers)
    Speed: 3111 statute miles per hour (5005 kilometers per hour), 4563 feet per second (1391 meters per second)
    Time: approximately 2 minutes after lift-off

MAIN ENGINE CUT-OFF
> Height: 69 nautical miles (122 kilometers)
> Speed: 17,500 miles per hour (28,158 kilometers per hour), 25,668 feet per second (7823 meters per second)
> Time: approximately 8 minutes after lift-off

ORBITAL OPERATIONS
> Height: 100 to 600 nautical miles (182.5 to 1095 kilometers)
> Speed: 17,181 miles per hour (27,644 kilometers per hour), 25,200 feet per second (7681 meters per second)
> Duration: 7 to 30 days

ATMOSPHERIC REENTRY
> Height: 76 nautical miles (140.7 kilometers)
> Speed: 16,629 miles per hour (26,756 kilometers per hour)

LANDING
> Cross range (right and left distance from runway): 1000 statute miles (1609 kilometers)
> Speed: 215 miles per hour (346 kilometers per hour)

Milestones in Space Exploration

October 4, 1957: Sputnik I (Soviet Union) was the first successful man-made object to orbit the earth—Dawn of the Space Age.

January 31, 1958: Explorer I (U.S.) became the first American unmanned spacecraft to successfully orbit the earth.

April 12, 1961: Vostok I (Soviet Union) successfully carried the first cosmonaut, Yuri A. Gagarin, into space; duration of flight—1 hour 48 minutes.

May 5, 1961: The first successful sub-orbital manned space flight by the United States. Alan B. Shepard, Jr., was the astronaut aboard Mercury-Redstone 3; duration of flight—15 minutes.

February 20, 1962: John H. Glenn, Jr., became the first American astronaut to orbit the earth aboard Mercury-Atlas 6; duration of flight—4 hours 55 minutes.

June 16-19, 1963: Valentina V. Tereshkova (Soviet Union) was the first woman to journey into space aboard Vostok 6; duration of flight—70 hours 50 minutes.

March 18, 1965: First extravehicular activity from an orbiting spacecraft. Alexei A. Leonov (Soviet Union) was the cosmonaut; duration of walk in space—10 minutes.

December 4-18, 1965: Longest manned space flight to this date by astronauts Frank Borman and James A. Lovell, Jr. aboard Gemini 7 (U.S.); duration of flight—330 hours 35 minutes.

March 16-17, 1966: First docking of two orbiting spacecraft, Gemini 8 with Agena target rocket (U.S.). Astronauts aboard Gemini 8, Neil A. Armstrong and David R. Scott; duration of flight—10 hours 41 minutes.

December 21-27, 1968: Aboard Apollo 8 (U.S.), astronauts Frank Borman, James A. Lovell, Jr., and William A. Anders were the first to orbit the moon; first to break free of earth's gravitational force. Duration of flight—147 hours 11 minutes.

July 16-24, 1969: First manned landing on lunar surface and safe return to earth; first return of lunar rock and soil samples to earth and manned deployment of experiments on moon's surface. Neil A. Armstrong, Michael Collins, and Edwin E. Aldrin, Jr., aboard Apollo 11 (U.S.) accomplished the historic mission. Duration of space journey—195 hours 19 minutes.

November 14-24, 1969: Second manned lunar landing accomplished. Moon's surface further explored and astronauts retrieved parts of Surveyor III spacecraft which landed in Ocean of Storms, April 19, 1967. U.S. astronauts who made the journey were Charles Conrad, Jr., Richard F. Gordon, Jr., and Alan L. Bean aboard Apollo 12. Duration of lunar trip—244 hours 36 minutes.

June 17, 1970: Longest manned space flight to this date made by Soviet cosmonauts Adrian G. Nikolayev and Vitaly I. Sevastianov aboard Soyuz 9. Duration of record journey—424 hours 59 minutes.

January 31-February 9, 1971: Third successful U.S. lunar landing achieved by astronauts Alan B. Shepard, Jr., Stuart A. Roosa, and Edgar D. Mitchell aboard Apollo 14. Duration of mission—216 hours 2 minutes.

July 26-August 7, 1971: Fourth successful U.S. lunar landing accomplished by astronauts David R. Scott, Alfred M. Worden, and James B. Irwin aboard Apollo 15. Lunar rover vehicle was carried on journey. Duration of mission —295 hours 12 minutes.

April 16-27, 1972: Fifth manned lunar landing with lunar rover vehicle (U.S.). Moon journey successfully accomplished on Apollo 16 by astronauts John W. Young, Charles M. Duke, Jr., and Thomas K. Mattingly II. Duration of moon flight—265 hours 51 minutes.

December 7-19, 1972: Sixth and final U.S. lunar landing with lunar rover vehicle. Astronauts Eugene A. Cernan, Harrison H. Schmitt, and Ronald E. Evans made the round-trip moon voyage aboard Apollo 17. Duration of journey— 301 hours 52 minutes.

May 25-June 22, 1973: American astronauts Charles Conrad, Jr., Joseph P. Kerwin, and Paul J. Weitz aboard spacecraft Skylab 2 docked with Skylab 1 (space station) for 28 days. Duration of mission—672 hours 50 minutes.

July 15-24, 1975: American astronauts Thomas P. Stafford. Donald K. Slayton, and Vance D. Brand docked their Apollo spacecraft with Soviet spacecraft Soyuz 19 with cosmonauts Aleksei A. Leonov and Valery N. Kubasov aboard. Duration of mission—217 hours 28 minutes.

August 20, 1975: An unmanned Viking 1 space probe (U.S.) was launched toward Mars. It orbited the planet, and then landed on July 20, 1976. The automated spacecraft took photos that were relayed back to earth and for eighteen months conducted computer-controlled experiments, the results of which were also transmitted to earth.

August 12, 1977: The Space Shuttle Enterprise (U.S.), a revolutionary spacecraft, made its first manned free flight with astronauts Fred W. Haise, Jr. and C. Gordon Fullerton aboard after release from a carrier aircraft. Duration of the free flight—5 minutes 23 seconds.

Glossary

Air lock—A chamber that can be made airtight to provide a passageway between two places having different pressure, as between two spacecraft docked to one another or between a spacecraft and a space station.

Avionics—A coined word from *avi*ation and electr*onics*. It concerns the application of electronics to the operation and control of conventional aircraft and spacecraft.

Deploy—To release or to open as with the manipulator arm aboard the Space Shuttle.

Elevons—A coined word from *elev*ator and ailer*on*, control surfaces of an aircraft. Elevons combine the functions of both. They are only effective on a conventional airplane or a spacecraft operating in the atmosphere.

Escape velocity—The speed needed by a spacecraft to soar beyond the pull of earth's gravity to attain orbit.

Gimbal—A device that permits a unit mounted to it to tilt in any direction.

High orbit—The orbiting path of a spacecraft more than 600 miles above the earth.

Launch pad—A concrete or other hard surface on which a rocket is positioned for launch.

Low orbit—The orbiting path of a spacecraft that ranges from approximately 100 to 600 nautical miles (185.2 to 1111 kilometers) above the earth.

Orbit—The path traveled by a celestial body in its revolutions around another body. A spacecraft does the same around the earth, moon, or other body in the solar system.

Oxidizer—A chemical substance that generally but not always contains oxygen for supporting combustion of a fuel or propellant.

Payload—The revenue-producing cargo carried by an aircraft or spacecraft.

Pod—A detachable enclosure or housing covering an extended part of a spacecraft's equipment or cargo.

Pressurization—Introducing and maintaining air pressure within an occupied section of a spacecraft, or an astronaut's suit, that is higher than the external pressure surrounding it.

Propellant—A combination or mixture of a fuel (such as liquid hydrogen) and oxidant (liquid oxygen) used to produce combustion and thrust for propelling a rocket. Propellants may also be combined or mixed in a solid form.

Reentry—The return of a spacecraft, manned or unmanned, into the earth's atmosphere after a journey in the airless world of space.

Reinforced carbon-carbon—A heat resisting all-carbon material reinforced for strength and treated to resist oxidation during repeated high-heat reentries. At temperatures destructive to metals, reinforced carbon-carbon actually increases in strength.

Rendezvous—The meeting of two or more spacecraft in flight at a prearranged time and place. Also, a point in space at which such a meeting takes place.

Safing—Removing all flammable and explosive elements from a grounded Orbiter following its return from space and before maintenance work begins on it.

Speed brake—A movable flap, or plate, which when extended slows an aircraft's speed by increasing its aerodynamic drag or air resistance.

Thermal heating—Aerodynamic heating produced by supersonic or hypersonic travel through the atmosphere; friction heat produced on a fast-moving body through the atmosphere.

Thrust—The force produced by a jet, rocket, or other type of reaction engine. Thrust is measured in pounds (or metric newtons) to represent the static pound weight moved by the engine. *Vacuum thrust* is the force produced by a reaction power unit under vacuum conditions or where the surrounding pressure is below atmospheric pressure.

Vernier—The word derives from Pierre Vernier (1580-1637), a French mathematician. In spacecraft usage it refers to small rocket engines or gas nozzles mounted on the outside of the spacecraft's body which can be tilted by commands from a vehicle's manual or computer control system to correct the spacecraft's attitude or position.

Metric Conversion Table

	Multiply	*By*	*To Obtain*
DISTANCE:	Inches	2.54	Centimeters
	Feet	0.3048	Meters
	Meters	3.281	Feet
	Kilometers	3281	Feet
	Kilometers	0.6214	Statute Miles
	Statute Miles	1.6093	Kilometers
	Nautical Miles	1.852	Kilometers
	Nautical Miles	1.1508	Statute Miles
	Statute Miles	0.8689	Nautical Miles
	Statute Miles	1760	Yards
LIQUID MEASURE:	Gallons	3.785	Liters
	Liters	0.2642	Gallons
WEIGHT:	Pounds	0.4536	Kilograms
	Kilograms	2.205	Pounds
	Metric Ton	1000	Kilograms
	Short Ton	907.2	Kilograms
PRESSURE:	Pounds/Sq. Inch	70.31	Grams/Sq. Cm
THRUST:	Pounds	4.448	Newtons
	Newtons	0.225	Pounds
VELOCITY:	Feet/Sec.	0.3048	Meters/Sec.
	Meters/Sec.	3.381	Feet/Sec.
	Meters/Sec.	2.237	Statute mph
	Feet/Sec.	0.6818	Statute mph
	Feet/Sec.	0.5925	Nautical mph
	Statute mph	1.609	Kilometers/Hr.
	Nautical mph (knots)	1.852	Kilometers/Hr.
	Kilometers/Hr.	0.6214	Statute mph
VOLUME:	Cubic Feet	0.02832	Cubic Meters

Further Reading List

Barnaby, Captain Ralph S. *How To Make And Fly Paper Airplanes.* New York: Bantam Books, 1970 (paperback).

O'Neill, Gerald K. *The High Frontier.* New York: Bantam Books, 1978 (paperback).

Ross, Frank, Jr. *Flying Paper Airplane Models.* New York: Lothrop, Lee, and Shepard Company, 1975.

National Aeronautics and Space Administration, Scientific and Technical Information Office. *Skylab, Our First Space Station.* U.S. Government Printing Office, Washington, D.C. 20402, NASA SP-400:1977.

National Aeronautics and Space Administration, Scientific and Technical Information Office. *Space Settlements, A Design Study.* U.S. Government Printing Office, Washington, D.C. 20402, NASA SP-413:1977.

National Aeronautics and Space Administration, Scientific and Technical Information Office. *Space Shuttle.* U.S. Government Printing Office, Washington, D.C. 20402, NASA SP-407:1976.

Rockwell International Space Division, Office of Public Relations. *Space Shuttle Transportation System.* Downey, California, 90241:1978.

Index

indicates photograph

Agena target rocket, 86
Air Force, United States, 14-15
air lock, 32, 34*, 88
Aldrin, Edwin E., Jr., 16, 86
Anders, William A., 86
Apollo spacecraft, 16-17, 49, 86, 87
Apt, Captain Milburn, 15
arm, manipulator, 32, 35, 58, 67*, 69-70
Armstrong, Neil A., 16, 86
astronaut candidates, training of, 31*, 32*
astronauts, 16, 22-23, 26, 28-29, 29*, 34*, 72, 85, 86, 87
Atlantic Ocean, 42, 53, 55, 70
avionics, 34, 88

Bean, Alan L., 86
Boeing, 17
booster engines, main, 37-39, 38*, 39*, 56
booster engines, solid propellant, 37, 41-42, 42*, 53, 54*, 83
Borman, Frank, 85, 86
brakes, speed, 45, 47, 61, 89
Brand, Vance D., 87

carbon-carbon, reinforced, 49, 50*, 89

cargo bay, 35-36, 69, 83
Cernan, Eugene A., 87
Chinese, the, 12
Collins, Michael, 16, 86
communications equipment, 47
computer guidance equipment, 37-38, 61
Conrad, Charles, Jr., 86, 87
cosmonauts, 15, 85, 86
crew-passenger compartment, 28-30, 28*, 30*, 32-34, 33*
Crippen, Navy Commander Robert L., 26

Dryden Flight Research Center, 21, 23*
Duke, Charles M., Jr., 87

Edwards Air Force Base, 21
elevons, 43-44, 47, 53, 88
engines, 37-39, 38*, 39*, 41-44, 42*, 53, 54*, 56, 59*, 60, 83
 main booster, 37-39, 38*, 39*, 56
 orbital maneuvering, 37, 42-43, 59*, 60
 reaction control, 43-44
 solid-propellant booster, 37, 41-42, 42*, 53, 54*, 83

93

Enterprise (OV-101), 10*, 19-23, 20*, 21*, 23*, 24*, 25**, 26, 29*, 87
environmental control units, 34-35
escape velocity, 55-56, 58*, 88
European Space Agency, 70, 71*
Evans, Ronald E., 87
Explorer I, 15, 85

facts, Space Shuttle, 11, 27, 83-84
flight deck, 28-30, 32
Ford, President Gerald, 19
fuel cells, 46-47
Fullerton, Pilot C. Gordon, 22-23, 29*, 87

Gagarin, Yuri A., 15, 85
Galileo spacecraft, 70
galley, 32-34, 33*, 47
Gemini spacecraft, 85, 86
gimbaling, 39, 47, 88
Glenn, John H., Jr., 85
Goddard, Dr. Robert H., 12-13
Gordon, Richard F., Jr., 86
Grove, Sir William, 46

Haise, Commander Fred W., Jr., 22-23, 29*, 87
hatches, 32
HRSI (high-temperature reusable insulation), 49, 51

Indian Ocean, 41, 56
insulation, 40, 48-49, 50**, 51*, 51-52
Irwin, James B., 86

Kennedy Space Center, 22-23, 26, 41-42, 53, 54*, 55, 55*, 62, 62*, 63*, 64
Kerwin, Joseph P., 87

kitchen facilities, 32-34, 33*, 47
Kubasov, Valery N., 87

launch pad, 53, 54*, 88
Leonov, Alexei A., 85, 87
living quarters, 32-34, 47
Lockheed, 17-18
Lovell, James A., Jr., 85, 86
LRSI (low-temperature reusable insulation), 49, 51
lunar landing, 16, 86, 87

Marshall Space Flight Center, 26
Mattingly, Thomas K., II, 87
Mercury spacecraft, 49, 85
midsection (middle deck), 32-34, 33*
milestones, space, 85-87
mission specialist, 30-31, 30*, 33, 53, 58
Mitchell, Edgar D., 86
model, Orbiter, 77-78, 80, 82
    flying, 82
    general directions for making, 78
    making body of, 80, 82
    making wing of, 78, 80
    materials and tools for making, 77-78

NASA (National Aeronautics and Space Administration), 17-18, 21-22, 30, 72
Nikolayev, Adrian, G., 86
Nomex, 51-52

Oberth, Dr. Hermann, 13
O'Neill, Dr. G. K., 75
OV-101 (Enterprise), 10*, 19-23, 20*, 21*, 23*, 24*, 25**, 26, 29*, 87
OV-102, 20, 26
oven, cooking, 32, 34, 47

Pacific Ocean, 42, 70
payload, 11, 32, 47, 53, 58, 67*, 89
pod, 43, 89
power, electrical, 46-47
power, hydraulic, 47
power stations, space solar, 72-73, 73*, 74*, 75
Princeton University, 75
propellant tank, external, 37, 40-42, 40*, 53, 54*, 55-56, 58*

rocket, history of, 12-13
Rockwell International Space Division, 17-19, 20*
Roosa, Stuart A., 86

safing, 63-64, 89
Sanger, Dr. Eugen, 13
satellites, 13, 15-16, 32, 65, 67*, 68-70, 68*, 88
 communications, 16, 65, 70
 high-orbiting, 68-70, 88
 low-orbiting, 65, 68, 88
 placing in geosynchronous orbit, 65, 70
 repairing, 65, 67*
 research, 16, 32
 weather observation, 16, 65
Schmitt, Harrison H., 87
Scott, David R., 86
Sevastianov, Vitaly I., 86
Shepard, Alan B., Jr., 85, 86
silica, 49, 51*, 52
Skylab, 72, 87
Skyrocket (D-558-2), 14
Slayton, Donald K., 87
solar power stations, space, 72-73, 73*, 74*, 75
Soyuz spacecraft, 86, 87
space colony, 73, 75-76
Spacelab, 70-72, 71*

Space Shuttle, artist's conception of, 57*, 60*, 62*, 64*
Space Transportation System, 11-12, 18-19, 26
"space tug," 68-70, 69*
Sputnik I, 15, 85
Stafford, Thomas P., 87
Starclipper, 17-18
"Star Trek," 19
statistics, Space Shuttle, 27, 83-84
supersonic aircraft, experimental rocket-propelled, 14-15
Surveyor III, 86

tail fin, vertical, 36, 37, 44-45
tail section, 36
tank, external propellant, 37, 40*, 40-42, 53, 54*, 55-56, 58*
Teleoperator Retrieval System, 68-70, 69*
telescope, space, 66*
television camera system, 35, 47
Tereshkova, Valentina V., 85
thermal protection system, 48-49, 50**, 51*, 51-52, 52*
transporter, 53, 56*
transport vehicle, Space Shuttle as, 65, 73, 75
Tsiolkovsky, K. E., 13
"tug, space," 68-70, 69*

V-2 rocket, 13
Vandenberg Air Force Base, 23, 41-42, 62, 64
Vehicle Assembly Building, 53, 54*, 55*, 64
velocity, escape, 55-56, 58*, 88
Vernier, Pierre, 44, 90
vernier thrusters, 44, 90
Viking 1, 87

von Braun, Dr. Wernher, 13
Vostok spacecraft, 15, 85

War of 1812, 12
Weitz, Paul J., 87
wing, double-delta, 27, 44
Worden, Alfred M., 86
Wright Brothers, 12

World War II, 13

X aircraft, 14-15

Yeager, United States Air Force Major Charles (Chuck), 14-15
Young, Navy Captain John W., 26, 87